THE ORIGINAL ENERGY THEORY

ELIER ENG

THE ORIGINAL ENERGY THEORY

PHOTOGENESIS

Copyright © 2011 by Elier Eng.

Library of Congress Control Number: 2011923906
ISBN:　　　　Hardcover　　　978-1-6176-4515-0
　　　　　　 Softcover　　　 978-1-6176-4513-6
　　　　　　 Ebook　　　　　978-1-6176-4514-3

Certificate of Registration:
TXu 1-323-787　Jun-25-07 in Spanish
TX　1-757-626　August-30-2007 English
TXu 1-741-991　June-2-2010 English
TXu 1-698-710　June-8-2010 Spanish
TXu 1-710-533　December 9, 2010

All rights reserved. No part of this book may be reproduced or transmitted in any form or by any means, electronic or mechanical, including photocopying, recording, or by any information storage and retrieval system, without permission in writing from the copyright owner.

Cover Disigned By
Elier Eng

This book was printed in the United States of America.

To order additional copies of this book, contact:
Palibrio
1663 Liberty Drive,
Suite 200
Bloomington, IN 47403
Tel: 877.407.5847
Fax: +1.812.355.1576
orders@palibrio.com

ACKNOWLEDGMENT

All the photos presented on the cover, back cover and inside the book were taken from NASA and NSSDC photos gallery. They permit the use of the images for any purpose.

I also acknowledge especially the Hubble telescope´s teams and all the scientists that have contributed so broadly to the knowledge on astrophysics, astronomy; contributed to the discovery of so diverse heavenly bodies; contributed to the discovery of the reality of the heaven.

MONEY has never been so well spent as in astronomy! I congratulate every participant for all the effort that have made.

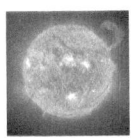
The Sun. Credit: NASA/ESA

The Earth **NASA** Photo **of** Planet **Earth** . . .

Quadruple Saturn Moons transit snapped by Hubble Credit: NASA, ESA and the Hubble heritage team. M. Wong
(STScl/uc Berkeley) and C. Go

Hubble Mosaic of the Majestic Sombrero Galaxy Credit: NASA/ESA and The Hubble Heritage Team STScl/AURA

Mars Closest Approach 2007 Credit: The Hubble Heritage Team STScl/AURA J. Bell (Cornell University) and M. Wolff (Space Science Institute, Boulder)

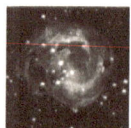
Light continue to echoes three years after stellar outburst. Credit: NASA, ESA and the Hubble heritage team. STScl/AURA

Hubble Space Telescope over Earth Credit: NASA/ESA

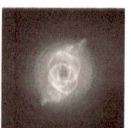
Dying star creates fantasy-like sculpture of gas and dust **Credit:** *ESA*, *NASA*, HEIC and The Hubble Heritage Team *STScI/AURA*)

The Heart of the Whirlpool Galaxy. Credit: NASA/ESA The Hubble Heritage Team STScl/AURA

Sunny Sided Up. Credit: Hubble Heritage Team AURA/STScl /NASA/ESA

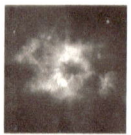
NGC 2440
Credit: *NASA/ESA* and The Hubble Heritage Team (*AURA/STScI*).

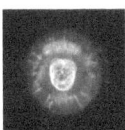
The Eskimo Nebula
Credit: *NASA*, *ESA*, Andrew Fruchter (*STScI*), and the ERO team (STScI + *ST-ECF*)

A New View of the Helix Nebula
Credit:
NASA, *ESA*, C.R. O'Dell (Vanderbilt University), and M. Meixner, P. McCullough, and G. Bacon (*Space Telescope Science Institute*

The Unique Red Rectangle. Credit: NASA/Hubble/ESA

Pinwheel Galaxy. Credit: European Space Agency/NASA

Pillar of Creation. Credit: Jeff Hester and Paul Scouwen (Arizona State University) /NASA/ESA

Galaxy NGC. Credit: NASA/ESA/Hubble Heritage Team

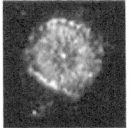
The Glowing Eye of Planetary Nebula NGC 6751
Credit: NASA/ESA/Hubble Heritage Team

Ring around supernova 1987 Credit: NASA/ESA/Hubble Heritage Team

The Earth-Base Best View of Mars. Credit: NASA/ESA/Hubble Heritage Team STScl/AURA

The Egg Nebula
Credit: *Raghvendra Sahai and John Trauger (JPL), the WFPC2 science team, and NASA/ESA*

The Egg Nebula
Credit: *Raghvendra Sahai and John Trauger (JPL), the WFPC2 science team, and NASA/ESA*

Out of this whirl: The Whirlpool Galaxy (M51) and companion galaxy. **Credit:** *NASA, ESA, S. Beckwith (STScI), and The Hubble Heritage Team*

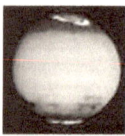
Comet Shoemaker-Levy 9 impact sites on Jupiter. Credit: The Hubble Heritage Team, NASA Hubble Photos Album

Spiral Galaxy NGC 4622 spins "backwards" Credit: NASA /ESA Hubble Photos Album

ELIER ENG

INDEX

ACKNOWLEDGMENT 5
PREPHASE 13
INTRODUCTION..17
COSMOLOGICAL MODEL 19
THE ORIGINAL ENERGY THEORY..28
PHOTOGEN...44
PHOTON 48
ELECTRON 52
SPACE TIME..55
TEMPERATURE..59
MATTER 64
PHOTOGRAVITON AND PHOTOGRAVITY 68
PHOTO GRAVITATION..72
ROTATION AND SPHERICITY..76
GALAXY SYSTEM..83
EXPANSION 87
BLACK HOLE THE RECYCLE SYSTEM................................92
ENERGY FIRST...100
LIGHT IS PHOTON PHOTON IS LIFE 108
ORIGIN OF LIFE...112
SYNGULARITY 119
FATE OF THE UNIVERSE 124
FATE OF HUMANITY 129
EASY, IMPOSSIBLE AT THE SAME TIME 133
CONCLUSION 139
REFERENCES...147

PREPHASE

The origin of the universe, even more, the origin of life has been the greatest question of human being, since the moment that humanity acquired the self perception faculty. It has been the restless work of scientists of all times. It has been the central theme of diverse religions. It has been the reason of terrific political, religious imprisonment, persecution, execution, even wars.

Under the dominion of the mysticism, uncountable Gods had been created to face up every inexplicable natural disaster phenomenon and challenges in live. Human ignorance undertakes bloods, goods, jewelry, virgins and life, to offer the angry Gods on purpose to have better climate, better harvest and better quality of live.

Ancient Gods had been abandoned while human kinds acquire knowledge. Religions had been changing their rules and concepts; even science and religion are still opposite.

The history has demonstrated that the dogmatic concepts, inflexible disciplines, repressive maneuvers almost had been ruled out and obligated to abandon under the irrefutable scientific discoveries.

The oldest written registries about astronomy were Aristotle, Ptolemy and San Agustin and maybe others had lost. They considered that the universe was formed spontaneously from nothingness and the universe is geocentric. This dominated idea was adopted by the Christian church during centuries.

Nevertheless being theologists, Copernicus, Galileo and Kepler corrected this point of view and proved scientifically the heliocentric theory. They realized that the universe was extended beyond the Sun and planets. Newton established the earliest mathematic and astrophysics' bases with his famous three physic laws. Einstein, Gamow LeMaître and others founded the scientific, contemporaneous astronomy.

The existence of God is still undisputable, beyond doubt! Would one day religion and science coincide in one point? God's power is energy; conscious scientific reasoning is energy. The path is already set up; maybe the differences reside in the inflexible reasoning.

The Original Energy as its name indicated is energy; energy is not touchable or tangible; we only could observe and detect its effects. The Original Energy theory might offer a lightening by attributing the origin of the universe, the origin of every existence and the origin of life to the most fundamental element the Photon which through Photogenesis guidance process constitutes everything. Photogenesis theory affirms nothing comes into existence by random unguided process.

Photogenesis theory does not deny the existence of God. Science should not be considered as devil's advocate! Science and scientists have been contributing to the better understanding, comprehension and illustration of the meaning **Heaven** which have been contributed to the progress, evolution and diffusion of theology. Copernicus, Kepler, Galileo, Newton, LeMaître, Hubble are live examples.

Be true or be fallacy the Original Energy theory, the author tries to open human's eyes and minds. We, human kind presume we are intelligent; we deserve everything on Earth, but we waste and destroyed everything that we touch or stand on; we kill each other to possess the Earth which barely is a tiny corpuscle in the immense universe; our meanness is as big as the space!

The atmosphere has protected us since our existence; nevertheless some people might be using electromagnetic energy to alter the atmosphere provoking massive destruction! Human kinds have not realized space is our heaven, space is our future and space is part of our home.

Hence, we should incorporate space development to our pacific coexistence. The first step should focus on the use of space energy by installing solar satellite or lunar solar station, concentrating solar energy and transfer the solar energy to the Earth and space industry. The utilization of solar energy could make the profit to keep investing in further space programs to exploit unlimited space resources: as the use of space electromagnetic energy, the use of comets, meteorites and make space journeys, etc. . . .

It is time of the space; it is time of concord and peace to conquer the space without war, the less possible alteration contamination of the environment!

It is time to go to the Heaven while we are alive!

INTRODUCTION

Theory could be conjecture, intuition, thought. It could be science fiction or real science. Hence, theory is something that could be or not could be. Theory depends on a specific space-time; it could be absolutely true in a determined moment, epoch, place, human mentality or grade of development and totally false in another moment, epoch, place, human mentality or grade of development. But something is sure: theory is the way to break through the mystery. Theory needs facts even if it has not been proved. Once it is proved theory would be reality.

The Original Energy Theory, the theory of Photongenesis is not a product of scientific experiment, investigation; as ancient time, it is a product of logical intuition. Since a long period of profound reflection and study, I hesitated to put one foot up to the prohibit heaven of astrophysics kingdom. I do not think I would be able to scientifically prove my theory in the rest of my life time. I don't even have such position, observatory or equipment of astrophysics science. But something urges me: Photon, the most fundamental element of light is the origin of the universe; moreover photon is an intelligent being, the origin of life! The name of Photongenesis derived from.

This theory has been published in a very abstract form in 2007 in my book "ECOS DE REFLEXIONES" in Spanish. But I always feel that it is not a fulfilling expression of what I really think; I hope this book could complete part of the commitment.

COSMOLOGICAL MODEL

A lot of theories have been exposed through the history; most of them believed that the universe derived from nothingness. It means, before the formation of the universe there were no matter, no energy, no space and neither time. This statement was accepted by the Catholic Church in the past century because it coincides with the creation, with the theology.

Since the 1880's scientists have observed the light that comes from far stars and galaxies is a red shift in the visual light spectrum, indicating they are receding from the Earth. This observation gave the impression that all the heavenly bodies depart from one point.

The cosmological model of the universe is derived from Albert Einstein General Relativity 1905, 1907, 1915's theory, extrapolating the expansion of the material universe backward: because galaxy and cluster of galaxies are receding now, under the action of the gravitational force, they should have been compressing more and more backward, up to a dimensionless point at a finite time. Temperature and pressure would elevate higher and higher while they were compacted.

Scientist concentrated their effort to prove this origin: Einstein 1910's singularity; George LeMaître 1927 postulation: the universe was formed from a huge explosion; Hubble 1929 confirmed the expansion of the universe; George Gamow 1930's dubbed the term of Big Bang theory and

Hoyle 1940's opposed the Big Bang theory but divulging the term Big Bang. This is the infinitely hot, dense primeval atom of the early phase of the universe; the most scientifically confirmed theory. The theory assumed that the Earth is located close to the gravitational center of the universe, which is the presumed site of the Big Bang explosion.

The Big Bang theory has been accepted as the COSMOLOGY MODEL base on:

1). Einstein's General Relativity theory, framework of the singularity where the gravitational force was infinitely strong, condensing all the heavenly matter, making it infinitely dense up to a primeval atom. This matter and energy primeval atom instantaneously came into being. Prior to that instant, nothing existed: no matter, any energy, any empty space or time;
2). the expansion of the universe which was observed since the 1880's and was confirmed by Edwin Hubble 1929 discovery. Hubble observed the galaxies and any material of the universe were red shift, receding from Earth which is the most direct evidence of the expansion. Friedman postulated the inflation of the newborn universe on 1920's;

   The most valuable of General Relativity and the discovery of the expansion of the universe demonstrated the universe is not static which even Einstein initially believed in, the reason from which he applied the Cosmic Constant;
3). The discovery of Penzias and Wilson in 1964 the presence of the cosmic microwaves background radiation (CBR), which is the afterglow of Big Bang cooling to a faint whisper. The blackbody spectrum of 2.7 Kelvin and the isotropic temperature of the universe are considered the strongest evidence of the Big Bang Theory; CBR was thermal when the universe was denser, hotter and optically opaque;
4). the presence of light chemical elements as hydrogen, helium as primeval dust, distributed homogeneously on the early universe, also has been attributed as a fundamental evidence of the Big Bang theory;
5). the Cosmological Principle consists in the provable isotropy and homogeneity distribution of matter and energy of the universe; the temperature all around the universe is 2.7 K. It is similar to the idea of Copernican Principal: from the Earth or in the solar system

there is no preferred observer or vantage point. The universe is not geocentric neither heliocentric; the physic laws are applicable everywhere and the individuality of the structures does not violate the universal laws;

6). the formations of large structures as planets, stars, galaxies were result of the growth of gravitational force derived from light primeval homogeneity; it means the new born universe was filled up with dust, the gravitational force emerged conglomerating the dust. On the action of gravitational force, the heavenly bodies were formed from anisotropic dust to planet, to stars, to solar systems to galaxy, to cluster of galaxies, from small to big;

7). nevertheless Einstein's General Relativity and Quantum Mechanics are irreconcilable; both form the most fundamental pillar of the cosmological model. General Relativity accounts for the gravitational force in extreme macrocosms, where the messenger graviton is the interaction particle. But there is no mass less spin-2 particle graviton, the quanta have not been found yet. On the other hand, the notion of smooth spacial geometry, the central principle of general relativity is destroyed by the violent fluctuation, the uncertainty which is the central feature of Quantum Mechanic.

Quantum Mechanic accounts in extreme microcosm atom with messenger subatomic particle: photon for electromagnetism interaction, Z W for weak force and gluon for strong force, respectively where the gravitational force is as weak as meaningless;

8). under the point of view of Cosmological Model, the action of gravitational force is considered essential in all aspect of the universe: the gravitational force conglomerates dust, atoms and mass to form stars; then gravitational force makes groups of stars to form galaxy: galaxies join together by the gravitational force to form cluster; then super cluster is formed by the action of gravitational force. That is why the universe could finish in singularity conglomerating all the mass together by the gravitational force;

9). the singularity has not happened because the presence of the cosmological constant which is "repulsive gravitational force" that maintains the material universe in a stable state. The Cosmological Constant is considered as the result of the presence of dark energy;

On the other hand, the existence of the space requires the existence of an intrinsic energy; it means, to maintain the space opened it requires energy. Because energy and matter are interchangeable, Big Bang theory considers energy should have gravitational effect! Then, it has been speculated the existence of dark energy to fill the space which has a negative repulsive gravitational effect that makes possible the accelerating expansion;

10). but the universe is expanding, the expansion is accelerating and the matter of the universe only constitutes less than 4%. It has speculated the presence of dark matter in vacuum to make possible for the gravitational force to hold all the matter of the universe together and not finish in grand endless fragmentation.

11). the fragmentations of big molecules to grand variety of subatomic particles culminated the idea of Big Bang. Nevertheless, the gravitational force is meaningless in subatomic, even in atomic level;

12). the fate of the universe depends on the density of the matter and the cosmic constant, the expanding rate of the heavenly bodies; it means depend on the attractive and the repulsive force of the gravitational force between masses.

The original idea of the Big Bang theory stated that the universe emerged from an immensely huge singular event which was an explosion that ever occurred about fourteen billion years ago. All the matter, atom, molecule, moons, planets, stars, galaxies and spacetime of the universe ware speeded forward and each time separate away. But it was not an explosion of matter that filled up the empty space of the universe. Instead, space-time expands after the explosion with the matter, making the distance between them increase. After that, the space on large scale is homogeneous and isotropic occupied by mass and vacuum.

Hubble observation of the red shifts galaxies, where the light emitted from them have shifted to longer wavelength evidenced the universe continued to expand, establishing the Hubble's law of the expansion: the further the galaxy, faster the acceleration.

Penzias and Wilson 1964 discovered the cosmic microwaves background radiation which is evenly isotropic in all directions. The CMR is consistent with the blackbody spectrum of 2.7 Kelvin. The dust anisotropy concentration made by the gravitational force contributed to the formation of atom, mass and the formation of all heavenly bodies, because the preponderance of the gravitational force over the others three forces.

Though, the Big Bang theory essential supports are:

The singularity reflexes a compressed material universe each time denser and hotter with its space and time concentrated in a primeval atom because the action of the gravitational force;

The Big Bang departed from an explosion of a dense hot material primeval atom from where all the matter of the universe is scatter radial away in all direction. Ironically not mass, not fragments of mass, neither heavenly bodies but just dust were encounter after the big explosion;

They are still joined together because the gravitational interaction;

The formation and evolution of the galaxies is believed that they derived from random condensation of anisotropic dust which form stars, then stars attract pulling them together by the gravitational force. It means the big structures as planets, moons, stars, galaxies, cluster of galaxies were formed from small dusts caused by gravitational force, not opposite generated from the nucleus of the galaxies.

During the stars nuclear fusion the electric repulsion of electrons balance the gravitational self-attractive force; but when the star is out of fuel, there is no more electric repulsion, the star collapses forming black hole causes by endless gravitational force;

Because energy and mass are interchangeable, the Big Bang theory considers the gravitational force governs the mass emptiness vacuum space;

Even the mass less photon is believed be attracted by the gravitational force in the lensing phenomenon and black hole where even the light could not escape.

Quantum Mechanic deals with subatomic particles which form atom and mass. The Cosmology Model has been scientifically accepted but it requires both quantum mechanics and gravitational force. Gravity should be taken in account. Hence Quantum Gravity theory was created.

From all these points we could conclude: the essential vertebral column of the Big Bang theory is in reality the mass and the mass dependent gravitational force. Any line of observation, explanation and conclusion arrives to mass dependent gravitational force. Hence, the Cosmological Model as Big Bang theory is gravitational mass dependent theories.

Because the Big Bang theory is gravitational mass dependent, it faces several problems:

I) when matter appears antimatter appear instantaneously annihilating each other. How could matter predominate over antimatter yielding the formation of the material universe and what force made the infinitely dense primeval atom to explode?

II) Galaxy formation: during the initial stage of the universe, the gravitational force was too weak and slow against to the expansion and inflation to permit the random formation of stars, galaxies from dusts. How could galaxies and the universe be created from unguided, random, homogeneity dust while the outward expansions, inflation pressure overcome the inward attractive gravitational force?

III) The radial emitted cosmic microwaves background radiations are homogeneously distributed and its temperature is almost spherically isotropic in the entire universe. In a hundred and eighty degree horizon the light (photons) travels toward opposite sides from the emission central point. How could they have the same density and temperature after they are separated twenty eight billion years? How could two galaxies in opposite side of the universe be identical?

IV) The density of matter and the rate of expansion are close making the universe almost flat. How could the universe be flat since everything was stretched radials?

V) The ignition of the Big Bang explosion: the primeval atom was the hardest material that ever existed in the universe. What provoked the ignition and what force could make such concentrated, compacted matter explode? Could the gravitational force by itself explode the primeval atom with one single explosion? Why the universe initiated with dust and light elements instead directly from primeval heavy matter to fragments of material heavenly bodies? How could an extremely dense tiny atom have spaces inside?

VI) the universe is homogeneous and isotropic, how could a clumping gravitational force that caused the singularity compressing all the heavenly bodies to an atom, could make such evenly distribution? In any kind of explosion matter should be speeded out unevenly on different velocity, in different size! It means the massive primeval atom should convert to fragments of molecules, blocks of mass, not just as hydrogen, helium light elements. Nevertheless the presence of light elements during the initial stage of the universe is one of the most important evidence of the Big Bang theory and the Cosmological Principle.

VII) Moreover, with what speed the fragments of the infinitely dense mass exploded out to form the universe in fraction of seconds while nothing, any material could travel faster than the speed of light?

Was the universe a globe expanding with matters and galaxies over the globe's surface which contain only empty vacuum with space-time inside the globe?

VIII) Matter and energy are interchangeable but it does mean never they are the same thing. People could burn up the entire forest of the planet in ten years to obtain energy but people need to wait more than four billion years to have another forest like the Amazon or Californian one to obtain such amount of energy again! Hence, through the singularity, all galaxies, vacuum space and time should be compressed. We would obtain an extremely solid and heavy "atom" or matter. Somehow it could be exchanged to energy and begin to form a new universe but the procedure, space, time and scenario would be completely different. It means base on gravitational mass dependent singularity the formation of new born universe should depart from dense, hot matter not from energy.

IX) Matter and energy are equivalent, they are potential time energy or kinetic time energy, their interchange procedure requires time, acceleration and complex transformation; it is not that easy as a formula on the paper! It is similar to what happen with the law of Newton, it could see the past, the present and even the future, but it might be impossible to see a broken egg to integrate never again!

X) There is too small amount of matter in the universe; there is not sufficient gravitational force to hold the entire material universe together. Dark matter is mass less even it has been named as matter, how could the false, missing matter has gravitational force to help the real matter? Then the Big Bang theory believes that the dark energy has repulsive gravitational force to cause the universe to expand. How could a perfectly proved contractive gravitational force be repulsive at the same time?

XI) Special Relativity theory stated nothing (any material) travels faster than the speed of light but two galaxies on the opposite side of the universe could be identical in consistence, even they are separated by hundred billion light years. General Relativity justifies this event, which contradiction difficultly could be explained.

XII) Randomness and spontaneity: under the action of gravitational force, the Big Bang theory attributed the formation of mass, heavenly bodies, even the formation of the universe to unguided randomness and spontaneity. How could the universe be so ordered and logic if it is true? Everything requires previous codes; how could the universe have an evolution procedure if it derived from randomness? If we could observe the brain it looks like completely disordered and complicated; fibers interlace everywhere but it is the most ordered and perfect computer, so is the universe.

XIII) The worse of all: according to Cosmological Model before the formation of the universe there were no matter, no energy, no space neither time; nothing existed. The universe comes to existence from nothingness. But according to the physic´s law: matter can be neither created nor destroyed. How could the universe come from unguided, nothingness, spontaneity?

XIV) There are some galaxies that are older than the Big Bang event; they were formed more than fourteen billion years ago. When and where they came from? Is there another old universe outside around the new universe?

On the other hand, the missing matter constitutes more than ninety per cent. What are they and where are they? What is the meaning of their existence?

XV) the Cosmological Model has supposed that the Earth is located close to the center where the Big Bang explosion occurred which should be the center of the universe, that is why matter and cosmic background radiation had scatted isotropic and homogeneously in all directions in the universe which constitutes the Cosmological Principle.

According to the Original Energy theory life emerged from the Earth because the universe is not geocentric neither heliocentric; Earth is localized in the comfort zone, far enough from the nucleus of the physical Sun, far enough from the nucleus of the Milky Way Galaxy and far enough from the nucleus of the universe. These imply far away from violent vibration, heat, explosions and far away where the gravitational force is extremely strong.

Every heavenly body revolves and has its specific orbit. Although the space it occupies only for instant, it is continually moving and changing. Consequently there is no fix position or time. We do not really know where the center of the universe could be, from where it was originated.

XVI) Every event, every observation, every conclusion has been attributed to gravitational force; even nobody knows exactly what forms gravitational force, what is graviton? Einstein´s General Theory of Relativity established Newton gravitational force as an illusion, there is no force at all! It affirms the curvature of the spacetime is what makes everything rotate; is what makes everything attract together. Gravitational force is inertia. How could Cosmological Model base on gravitational force, considering the force that govern everything in the universe and denies its existence at the same time?

XVII) the Big Bang theory affirms that the universe departed from a dense, hot, primeval atom product of the action of immeasurable strong gravitational force. The gravitational force compressed all the matter, stars, galaxies, space and time, elevating extreme temperature, causing the singularity. Here is the big confusion: there was a universe which finished in singularity previous the Big Bang event or the Big Bang event occurred from nothingness? If the Big Bang event appeared from nothingness, when the singularity compressing the entire material universe occurred?

Anyway the Big Bang theory only descript the initial phase of the universe; it does not descript why, how, what, when and where the universe was formed; it means, Big Bang is not the real origin of the universe.

THE ORIGINAL ENERGY THEORY

The Original Energy theory establishes that the universe derived from an embryonic energy formation which was cold, containing ultra energetic photons with the energy trillions of trillions of times stronger than the actual photon´s energy. *Through Photongenesis procedure, a mutual generation between photons and electrons, extreme energetic photons extended their wavelength transforming to less energetic photons.* They possessed all the codes capable to generate all existence being including the creation of life inside the universe.

It establishes that before the actual universe was formed, the Original Energy or the Energy O already existed, where the embryonic photons were conglomerated in a tiny compact, freeze energy formation. The Original Energy embryonic photons were extremely small. Under almost 0 K temperatures, photons were points like without rotation or movement, without electric or magnet vibration, without electromagnetic field around them. Hence, the Original Energy was one without distinction as kinetic or potential energy; neither as space time energy; there were no four forces distinction yet.

The Original Energy Theory is an energy based theory, which affirms that every heavenly body, object or energy formation possesses intrinsic ultra compact Original Energy inside the core, responsible of every generation, formation, development, evolution, transformation, rotation, action o reaction. Being photon the basic unit of the electromagnetic interaction, it implies that such creative energy was electromagnetic energy. It implies that it was not the mass and the mass dependent gravitational force from where the universe derived from.

Time existed as the eternity past, since the infinite past, belonging to others cyclic transformations.

Space existed as a Virtual Space included inside the compressed Original Energy, characterized by the absolute mass emptiness. Space-time was inherent inside the Original Energy, all in one energy concentration.

During this embryonic stage of the universe, mass was zero, space was zero, time was zero, even the temperature was close to zero. Everything begins from zero but the immense energy.

When the embryonic photons mature, the Original Energy through Photogenesis processes emitted messenger photons, precursors of the actual photons with infinite compact wavelength, capable to develop extreme high frequency. Shall we call them as photogens?

From photogens derived electron and positron which reacted converting to pairs of less energetic photons; photogens generated pairs of electron and positron which were converted to pairs of less energetic photogens again; from new photogens derived new electron, positron emerged, transforming to less energetic photons. They rotated, vibrated and polarized; the magnet bar was formed inducing the electric formation. The electromagnetic field was completed. This Photogenesis, a mutual creation between photons and electrons processes was repeated again and again.

When there were excessive amount of photogens inside the compact system, they vibrated intensely, the electric and magnetic components were extended and turned to perpendicular to each other. They commenced to unwind and stretch violently elevating extreme high temperature and pressure producing a huge explosion. Photons flied out faster than the speed of light!

If we analyze the electromagnetic spectrum the visible light is a narrow 1/1000 cm band, the rest of the spectrum is invisible. During the formation of the universe all the photons were extremely energetic extra gamma and ultra gamma rays, higher than the gamma rays of the normal electromagnetic spectrum. They were not visible! The universe was not visible until millions of years after it was born! There is no way to make photographs of that period.

From photongens derived electrons and positrons annihilating each other, producing a thermonuclear reaction. The universe appeared as an extremely hot fire ball of photogens and photons which energy was distributed isotropic and homogeneously. Then the universe was visible as a fire sphere.

The hot, infinitely dense, fold and compressed Original Energy were violently unfolded, unwind and spread. After series of explosions, the isotropic and homogeneity energy of the new born universe was altered. It was scattered establishing billions of multiples fire balls centers which in the future will convert to the nucleus of stars and galaxies. That was the period of Nucleogenesis processes when the formation of the nucleons of all the heavenly bodies took place.

The frequency of the photogens diminished, the wavelength extended progressively. The Original Energy forms the electromagnetic structural mesh of the whole universe.

The Photogenesis theory establishes that the universe was originated from the Original Energy; the photons gave origin to every existence in the universe, including life as vegetal and animal on Earth, from where the name of Photongenesis is given. The infinitely high frequency, compact wavelength and extremely small photons of the Original Energy were converted to lower energetic photons, while they were divided, stretched, unwind, extend the wavelength, amplitude progressively, constituting the force that inflated and expanded the universe.

These were the ways the universe was formed from infinite energetic compact photons:

1). Duplication of photogens; photogen is built by electron and positron they were separated converting to pair of less energetic photogens which converted to electrons and positrons again; they transformed, becoming less energetic photogens. These processes were repeated successively again and again. The electromagnetic field was formed, electric and magnetic components were turned to perpendicular to each other;
2). unfold, unwind, extend and scatter, photons flied out and travel faster than the actual speed of light;

3). electrons derived from photons, positron appeared annihilating each other; from annihilation between electrons and positrons derived new photons with less frequency and longer wavelength. New photons released new electrons; positrons appeared annihilating each other deriving a pair of new photons, new photons released new pair of electrons and positrons; after the annihilation new pair of photons appeared and so on repeatedly. The compact, dense energy was extending, unfolding, each time with longer and longer wavelength. This was and still is the reason of the expansion and inflation of the universe;
4). by ionizing the ultra and extreme high energetic photons (Energy O, Photogen) transformed to plasmatic state releasing electrons, becoming to lower energetic photons;
5). in the radiation dominated universe, during the inflation period, the universe expanded faster than the speed of light and cooled down quickly. The stretched photons caused the space to expand, which equivalent to cause the wavelength to extend and the frequency decreased. Original Energy and photogens occupied the universe as a fire ball during long period while they stretch. This period was characterized by hot, dense and opaque; photons were not visible yet;
6). the early formation time, represent the most energetic epoch of the entire evolution of the universe. It means the actual photons come from extremely higher frequency which were photogens that formed the nucleus of every heavenly body;
7). from photogens derived less energetic, common photons with all ranges of electromagnetic waves. By releasing electrons ultra and extra gamma rays converted to common gamma rays; by releasing electrons gamma rays were converted to X rays; by releasing electrons, X rays converted to ultraviolet rays; ultraviolet to violet rays, to all the visible light colors rays, then to infrared rays, then to microwaves rays and so on, down to radio rays. New, less energetic photons were formed;
8). photogen is the transformer; it has a dynamic activity to be energy or matter. In a further stage, photogens converted to all kind of subatomic particles, transferring its energy. In any action and reaction less energetic photons were released by releasing electrons;
9). once galaxies and stars were formed, matter decays, atoms give off radiation. Electrons jump from more energetic orbit to less energetic orbit releasing less energetic photons;

10). inside the nucleus of the galaxy or star the nuclear fusions release photons from higher energetic photogens to lower energetic level of common photons. Hence, the entire universe is made up from the transaction of electrons and photons and its final product from disintegration always are photons. There is no other secret!

11). the early formation time also represent the hottest epoch which was the condition required for the formation of heaviest elements and mass; these extremely hot scenario does not repeat until the formation of supernova; while the radiation of photons got cooler, less energetic the photons become. The Photongenesis theory states most of the heaviest elements are localized in the nucleus of the heavenly bodies, most of them were formed during the early hottest epoch when the Nucleogenesis of the universe occurred. Posterior, supernovas contribute to the formation of the heaviest elements that found in the outer layers of the heavenly bodies and vacuum space.

Original Energy theory states that all activities of the universe derive from energy. Fusion is realized in the core of our Sun, where high energetic photogens release radiations with positive pressure, which overcomes the attractive gravitational force. This radiation pressure also defeats the well known electron-positron annihilation, releasing less energetic photons. It means electron-positron annihilation is a coalition releasing light and solar radiation; it is just part of the transformation from the core of the stars, not a fatal final.

During the energy dominate epoch, during the high energetic photons fire ball epoch, gravitational force did not exist since matter did not exist, and could not exist, because the extremely high temperature could burn it up as soon as it appeared. Once the temperature diminished enough and the subatomic particles formed, gravitational force emerged, but it was too weak, too late and too slow. The positive radiations pressures overcome. Hence, that was the reason matter overcome antimatter; also because the radiations outward force was stronger than the weak gravitational force that was the reason the new born universe could form and not be collapsed by the gravitational force!

During the formation of the new born universe, the universe consisted only with light and heat, outside of the extremely hot, dense, compressed fire ball there was no pressure; there was no any resistance. There was no mass; there was no gravitational force inside to counteract the positive radiation pressure. Photons inflated violently the universe with the speed faster than the speed of light.

Other aspect is the rotation and sphericity of every object and the entire universe. The Original Energy Theory attributed the homogeneity and the Cosmologic Principle to the effect of rotation of the Energy O. The rotational force revolves and mixes, making most of the object spherical or elliptical and is the cause of the homogeneity of matter and energy. Also because of the revolving effect the temperature is isotropic. This point is also favorable to the Original Energy Theory. The initial explosion was radial; hence the universe was sphere, expanding with space and time filled up by photons. The forces that expanded and inflated the universe were the extension and unfold kinetic energy of photogens and photons.

The rotation of the atoms, stars, planets, moons, galaxies and of the universe is made by an intrinsic force, primarily by the electromagnetic force of the Original Energy which forms the axis of every heavenly body, not by other external force or factor. Photons and electrons spin with intrinsic energy all the time which makes possible the rotation of the axis of every object and heavenly body. Outside of our universe there was no space-time, how could space-time warps making our universe to rotate? How the square space-time fabric makes the heavenly body keep in specific circular orbit? Why the Earth does not drop down to the Sun while circulating around a deep bowl of the space-time cause by the Sun? The square space-time fabric is not smooth, why the heavenly bodies not rotate by jumping? In reality, all the actions and transformations of the universe happen because the real action is realized by the intrinsic Original Energy that every object possesses.

The most extraordinary part of the Original Energy, it exists in the entire nucleus of: galaxies, stars, planets, moons, atoms, cells, seeds, ovum, spermatozoid, eggs, and neuronal system; it forms the nucleus and axis of the universe as well. It posses the codes of every existence, making everything ordered, logical and organized. Codes are made up with photons of different ranges of electromagnetic wavelength and frequencies.

The tiny, the tiniest particle, the electron, spins all the time, with its intrinsic energy, that is only possible with tiny amount of Energy O. the space-time fabric could never make it spin. Space-time is built up with Energy, the energy that irradiates from the core of every heavenly body.

Original Energy forms the structural mesh of the universe; it is the intrinsic energy that forms the space; it possesses codes and messengers with which leads and transforms the universe. Its action is instantaneous, undetectable with conventional methods, without locality or temperature and is faster than the speed of light. If we understand this point, it should not be difficult to understand why the universe was formed so fast!

The Original Energy rules and takes control of the nucleosynthesis, the formation of the stars, galaxies, clusters of galaxies, black holes recycle process, spacing of matter, equilibrium distribution between matter and energy.

When the energy is exhausted, for instant, in a star or galaxy, the nucleus and the axis remain revolving violently gathering remnants energy, mass, ashes, light and radiations until it converts itself in a black hole. The black hole recycles everything and converts all of them into atoms, then to subatomic particles, then to photons and electrons. Temperature would elevate trillions of degree until the black hole explodes, releasing enormous energy. The common photons are compressed, energized by electrons becoming to photogens, to higher energy frequencies and more compact wavelength. In a further stage it could converted to a supernova. New stars are formed again or the energy just incorporates to the universe.

Hence the Original Energy is an integrate energy that could stretch unwind convert to common photons or opposite compress rewind the common photons converting to higher energetic photons. It is formed by three levels of waves, should we name them as:

Ultra gamma rays or O rays: they are extremely small photons, with most compact wavelength radiation which form the Original Energy. They could convert to extra gamma rays or any range of electromagnetic rays conserving the codes that could generate or transform. They could be everywhere in the universe;

The extra gamma rays or E rays: they are extra short wavelength which constitutes the transformer photogens that could transform to energy/mass or mass/energy, the most likely to particles;

The common range photons or C photons: derive from photogens which constitute the electromagnetic spectrum rays from radio, microwaves, infrared, visible color light rays, ultraviolet, up to X ray and gamma rays.

In all of them the fundamental elements are photons. The first and second types reside in the nucleus of any star or galaxy, space, black hole and supernova. The third type photons in solar system, space and anywhere of the universe.

The Original Energy is constituted with three parts: Energy O, photogens and regular photons of the electromagnetic spectrum. Any of these entities could transform to potential space-time energy or kinetic time energy and the energy is conserved. Photon by itself is built up by kinetic and potential energy.

Photon has wave particle like behavior, longer the wavelength, more wave like its behavior; shorter the wavelength more particle like its behavior. The Original Energy theory states this behavior is because higher the frequency of the photon smaller is its size; higher the frequency more compact the wavelength. Photon could twist, rewind, swirl, and squeeze, converting itself in a particle like ball.

Photon's waves-particle duality behavior not only exists in their structure, it also reflexes in their functional behavior in all three levels of photons as well. For instant, kinetic and potential time energy; magnet and electric components of the electromagnetic wave and field; crest and trough propagation, photosynthesis processes, mass and energy transformation, binary formation of stars or galaxies, etc. . . .

The most important characteristic of the Original Energy is the intelligence center, possessing all kind of information; it is highly organized and structured. It rules every activity of the universe.

Original Energy comprises all the codes that determine born, life, dead, recycle, surrounding condition and duration time of the existence of everything that existed, exists or would exist. Even things should never exist. Just like a seed, an ovum that contain all information about the components, space and time of a plant, an animal or any kind of object.

The Original Energy forms the energetic axis and nucleon of any structure and is the force of rotation of any object, since subatomic particles, atoms, planet, star, galaxies, cluster of galaxies, up to the entire universe.

It forms the structural mesh of the universe; through photogens fills up the entire space constituting the support of every structure. Its action is instantaneous with no locality because it is distributed everywhere of the universe as a web. The structural mesh does not the one that travels, the heavenly bodies and the light are the ones that travel. The interconnectedness between the web, subatomic particles, photons, demonstrated the no locality and instantaneous correlation of the whole cosmos, regardless of the separation, thanks the Energy O. The Original Energy mesh is an integrated network; it is sensitive to every change.

The Original Energy is the origin of photogens and photons where the Photongenesis process occurs. It is the medium through where the photon, the most fundamental element of the universe travels.

The Energy O is a revolving energy, anything derived from it spins and generate electric charge and magnet, consequently forms part of the electromagnetic field.

Almost all the astrophysics theories are base on matter dependent gravitational force. The Big Bang initialed with an extremely hot, dense, solid primeval atom, mass of the entire universe that contracted to a point, which was the singularity. Hence, once the explosion or Big Bang occurred, large, solid fragments of material should be found. The homogeneity and isotropy should not happen. Once the inflation occurred, the stars, the galaxies should appear and might have a similar arrangement to the precedent Universe! But at the beginning, such mass was so immeasurable dense and hard that made an unimaginable strong gravitational force that there was nothing could separate and scatter it.

The Original Energy Theory is base on infinitely dense, cold, folded wave energy, the electromagnetic energy that generates and rules the universe. The wavelength of the photons has extended and the energy has diminished during around twenty billion years. It means at the beginning of the formation of the universe the wavelength was extremely compacted, folded and the frequency of the energy unimaginably high. But once the Energy O got free, the waves extended and unfolded violently. That is the reason the universe could form, inflate, and expand faster than the speed of light. That is the reason the universe keeps expanding. In vacuum only Original Energy and photogens and their ancestral vestige existed.

These are transcendental factors that prove the veracity of Original Energy theory:

1). the possibility of the existence of an embryonic energetic formation previous to the event of explosion. The exclusive existence of ultra energetic photons, electrons and the extremely high temperature in the fire ball after the explosion of the initial stage of the universe, where any matter could not exist, neither its formation could be possible. Consequently, the gravitational force did not exist;

2). the exclusive existence of photogens and photons constituting the electromagnetic structural mesh and the electromagnetic interaction at the initial stage of the universe. The universe was not formed by randomness from dust to heavenly bodies as the result of the attractive gravitational force; photons made and still making an ordered structure;

3). the violent rapidity formation of the universe. The absent of gravitational contractive force inside the fire ball; the absent of pressure outside the fire ball permitting the outward inflation and expansion. The extension of the wavelength of photons; the extreme elevate electric outward pressure inside the fire ball made possible the inflation which determined the fast formation making possible the formation of the universe and prevented it to collapsed;

4). during the new born universe epoch the universe was characterize by absolute matter emptiness; the universe was a big fire ball with extremely high temperature. While the universe expanded the heat was transmitted exclusively by radiation, not by convection or conduction, because there were no any matter particles fluids to transmit that heat. The persistence of cosmic microwaves background radiation, the unique thermal vestige of the new born universe proved it and not proved the existence of any chemical elements;

5). the progressive extension of the wavelength, the diminish level of energy and lower frequency of the photons. Photons redshift is directly proportional to the distance that they traveled. These are one of the most outstanding factors that prove the Original Energy theory;

6). lighter heavenly bodies have been rotating around the stars, stars around the galaxies, galaxies around the universe's axis during more than fourteen billion years, the homogeneous and isotropy distribution of them and of the temperature are caused by the rotational effect of the axis of the Original Energy. The contractive gravitational force only could cause anisotropy and heterogeneity;

7). during the initial stage of the formation of the universe there were no matter exist but photons and electrons, then the subatomic particles were formed. Proton and neutron emerged. In a further stage, temperature and pressure diminished, gluon enters in action, electron bound proton of the nucleus, and light chemical elements were formed. This is the reason of the abundant existence of light elements as hydrogen, helium during that initial epoch; it means the light chemical elements were formed from even smaller subatomic particles not from the pulverization of the extreme condense mass of stars, galaxies and clusters of galaxies that come from an uncertain singularity. If the universe was formed from singularity, from extreme dense matter, from an extremely heavy solid primeval atom, after it exploded it should consisted of all kind of atoms, composites, heavy fragments of mass, even parts of celestial bodies. After the Big Bang the universe should not be form exclusively of light elements. Homogeneity and isotropy should not exist after the Big Bang event. Worst of all: if the universe was formed spontaneously from nothingness, what made the singularity?

The Photongenesis theory establishes: atoms, light chemical elements, composites and mass were and still are formed by Nucleosynthesis processes in the nucleus of the stars and galaxies, posterior to the Nucleogenesis formation;

8). proton is positive charge; neutron is identical to proton but without electric charge. Protons could bind with electrons to form atoms because electrons are negative charged. All the chemical reactions are carrying out with the exchange and combination of electrons. It means, atoms formation, composite formation, mass formation, solar system formation, even clusters formation are electric magnetic interaction. Like charges repel, unlike charges attract and the attractive electric force is inversely proportional to the distance. Consequently, the real force that binds every existence is the electromagnetic force not the gravitational interaction. Moreover, the Original Energy forms the core and axis of every existence that form photons, electrons, and the electric and magnetic attraction. It implies that the attractive interaction between heavenly bodies or objects does not need to consider a gigantic mass as a point like in their center as the Newtonian gravitational force needs it; nor the wrapped curvature of the spacetime as Einstein General Relativity needs it. It is the intrinsic electric magnetic Energy that makes everything attracts, rotates or separates; it implies that the attraction or repulsion is not made by gravitational force;

9). the expansion of the universe derives from the extension of the Original Energy structural mesh and wavelength of the photons, not by the expansion of the space itself; the contractive gravitational force could never cause the expansion;

10). the most trustworthy evidence that redshift is result of expansion of the universe is the thermal spectrum of cosmic microwave radiation which resulted from the propagation of the Original Energy fire ball radiation, vestige of the initial Energy;

11). the existence of extremely high frequency rays burst, belonging to the oldest and furthers galaxies which are trillions of trillions of eV, confirm the existences of ultra and extra energetic photons. They are as important as the existence of the background microwaves radiations;

12). the continue transformation of energy to mater, matter to energy which depends on completely on the transformation of kinetic time energy and potential space time energy which form the two components of photons; it means the transformation depend on photons energy not the action of the gravitational force. Gravitational force does not make chemical reaction, neither transformation; it is a static force;

13). every heavenly body has an energysphere that emanates from the nucleon and reach several times its diameter. Same phenomenon occurs outside the material visible universe; spacetime is part and is formed by the energysphere. Consequently any action of the curve spacetime in reality is the action of the energysphere of the heavenly body;

14). when old stars or galaxies exhaust of energy, they transform to black hole; the remnant ashes, material, light, radiation are transformed to photons and electrons by the black hole which constitute the recycle system.

Only the transformation between potential energy and kinetic energy of photons could make the existence of black hole possible. It is not the endless gravitational force that formed the recycle system of black hole. The formation of black hole primordially is a procedure of transformation of potential time energy to kinetic time energy! It is a transformation of matter to photons and electrons, a photonsynthesis procedure, not a singularity;

15). the eternal transformation of the Original Energy makes possible the endless transformation of the universe. Photon possesses kinetic energy and potential energy, the interaction of these two energies constitutes the transformation of the entire universe. When the electric and magnet components are perpendicular, the photon travels with the speed of light; photon possesses the total kinetic energy. When the universe was a fire ball the kinetic energy was at the highest level of energy.

When the magnetic and electric components are compact and energized extremely by electrons, photons are compressed and put on rest; photon's electric and magnetic field reduce up to a point. All the kinetic energy is transformed to Original Energy, inactivating the electromagnetic field. This was the embryonic state of the universe or the plenitude of the formation of celestial's bodies, before they decay. In this moment, energy and mass are equivalent E=m, just because they are made by the same ingredient which are photons without acceleration;

16). the probable longevity, fate of the universe, because the abundant existence of energy in the universe that still could stretch, extend transform and recycle;

17). the uncertainty cause by the transformer photogens. Photogens have the faculty to appear or disappear transforming either way to energy or matter. Because the electromagnetic interaction is the force of energy or matter;

18). in vacuum space, as in the Earth has been encountered highly energetic particles that appear spontaneously in empty detectors which have not been elucidated, how they entered there. This could prove the existence of Original Energy that is distributed everywhere and could transform to photons and subatomic particles;

19). while the gravitational interaction is adequate for macrocosms and General relativity, it is meaningless in microcosms and quantum mechanics. Contrarily, the Original Energy affirms photons and electrons are the most fundamental constituents of both energy and matter. Hence, electromagnetism account in both microcosms and macrocosm; it means photons and electrons of the Original Energy account in microcosm and macrocosms.

20). if we analyze the volume of an atom, the nucleon formed by proton and neutron is extremely small; the electron that rotates around is even smaller. They, proton, neutron and electrons occupy an insignificant space. The rotation of the electron is electric charge creating electric field and the electric field induces the formation of magnetic field; together form the electromagnetic field. The rest of the vast volume of an atom is empty space fill up with electromagnetic energy! This phenomenon occurs in the entire universe. It means the principal ingredient of the universe is electromagnetic energy built by photons!

The Photongenesis theory states the matter and the energy of an atom form a unity not just the matter;

Earth and its geoenergysphere form a unity; the Sun and the heliosphere, all the energysphere form a unity; any heavenly body and its energysphere form a unity. Consequently, mass and spacetime form a unity, together form the gravitational field, not just mass as Newtonian theory stated neither just the curved spacetime as General Relativity stated; spacetime is formed by the electromagnetic field, the energysphere that come from the core. Nevertheless, the heavenly body does not curve the energysphere; it forms the energysphere by consuming transforming the potential energy to kinetic energy;

21). by the same reason any heavenly body is bigger than it apparently looks like; the universe is a lot bigger than the observable galaxies have reach. Any old galaxy is enclosed inside the energysphere of the entire universe;
22). Photons of any frequency travel with the speed of light in vacuum. When photons enter to a medium with different density, they undergo to reflection, refraction, diffraction, interference or polarize and slow down. When they leave the medium, they travel with the speed of light again. Electron spins billions of times every second. These indicate that photons or electron possess intrinsic energy which is the Original Energy and not depend on the warp of space time. The lensing phenomenon happens when the photons of the far located star or galaxy penetrate the energysphere of a galaxy or the Sun and change their velocity; it is not because the gravitational force bent the light. It means lensing phenomenon is not cause by the action of the gravitational force that bent the mass less photons which could not escape;
23). the Photongenesis theory states that the universe derived from an extremely high energy, trillion and trillion stronger than the actual gamma rays of the electromagnetic spectrum. The ultra gamma bursts that come from old galaxies confirm this statement. These ultra energetic gamma rays have demonstrated even more, they could transform to less energetic up to the common electromagnetic spectrum rays;

24). the Photongenesis theory has postulated that the universe was formed, expanded, inflated, and keep expanding because the ultra and extra compact waves of the Original Energy photons have uncoiled, extended their wavelength. The theory also forecasts and affirms the existence and appearance of ultra and extra week variety of photons; these new photons would keep making possible the evolution of every existence in the universe. The constant transformation of the Original Energy makes possible the continue changes, the ordered renovation of the universe.

The fate of the universe does not depend on the rate of the expansion and the density of the existent matter. It means, it does not depend on the Cosmic Constant that the same Einstein considered the worse error of his life; error that the Cosmological Model and Big Bang theory insist to hold;

25). the visible light constitutes only a narrow portion of the electromagnetic spectrum, the rest of photons are invisible. The higher frequency photons, higher than the gamma rays photons are invisible and even undetectable; they form all kind of dark existents as dark energy and dark matter.

All these confirmed the veracity of the Original Energy instead others theories. Hence, dark energy, dark matter, singularity, the entire Big Bang, Cosmological Model theory might need revision.

The Big Bang theory departs from mass while Photogenesis theory departs from energy. It has been established and it is well known that mass and energy could be interchangeable, are equivalent. Hence both theories could be considered the same. Yes, that is almost true! They are almost the same. Photons possess kinetic and potential energy: when photons travel with the speed of light, it possesses only kinetic energy; when photon is in rest photon energy is potential energy. But it could take twenty billion years or more to make possible the material universe (potential energy) converts to a fire ball (kinetic energy) as the initial universe! It could take four billions years for the formation of the Amazon and Californian forest and only one month to burn them up! Hence, matter and energy could be radically different in which the Factor Time is decisive! That is why the transformation is cyclic because fate is time. We should not forget: to make possible matter equal to energy it requires acceleration and for acceleration it requires time.

The Photongenesis theory is energy based theory where the energy always is conserved. It means, the Original Energy of the universe would always transform cyclically and persist until the eternity.

Under the guidance of the Original Energy, Photons gave origin to every existence of the universe. The origin of all heavenly bodies, all matter and the origin of life are attributed to the most fundamental element the Photon under the Photongenesis process. Photogens contain all the codes of every existence in the universe, which are uncountable varieties of different combinations of electromagnetic waves. Photongenesis theory affirms nothing come into existence by random unguided process. Everything, the entire universe is so ordered, so logic, so organized, so natural because the guidance of the Original Energy.

The Original Energy as its name indicated is energy; it is not touchable, visible or tangible; we only could observe and detect its effects. God's power is energy, it is not visible, touchable neither tangible.

Photongenesis theory does not deny the existence of God. Science should not be considered as devil's advocate! Science has been contributing to the better understanding of the meaning **Heaven;** science has made monumental progress, better understanding, even evolution on theology. Copernicus, Galileo, Newton, LeMaître, Hubble . . . are live examples. Yet, Copernicus, Galileo, LeMaître were theologies that contributed to this evolution.

PHOTOGEN

At the beginning, during the embryonic period when time was zero, the Original Energy emitted messenger photons which possessed virtual extra high frequencies, their energy was trillion trillions of times higher than the now existing common spectrum of photons possess. But they were at rest, small point like, with wavelength extremely compact, without electromagnetic field. They possess the faculty of generation, duplication and transformation; it means the faculty to generate diverse spectrum of photons. They contained the codes of every being, the entire value of space and a complete cycle of time. Shall we name them as "Photogens"?

Photogens commenced a division/duplication process inside of their energy formation, when they were too crowded they vibrated violently. The Original Energy commenced to spin and rotate. The electromagnetic field emerged. From photogens derived electrons and positrons which reacted. They heated up, elevating the temperature and pressure extremely. Huge explosion overcome.

The photogens were fragmented and scattered in radial form, with a speed higher than the actual speed of light. Photons begun to extend, stretch their waves and the frequency diminish. Nothing could exist yet under that high pressure, high temperature and violent vibration but energy of photogens. The four forces were the same stem: the electromagnetic force of the Original Energy.

The new born universe appeared as a revolving, big spherical fire ball, constituted exclusively with isotropic plasma of Original Energy and photogens, creating a new time, a new space and a new independent universe in the immense cosmos.

In a further step photogens generated pairs of electron and positron; from the annihilation emerged pairs of photogens. Photogens and pairs of electrons with positrons were generating each other alternately and successively, trigger violent reactions with extremely high temperature and pressure. The new born universe was inflated and expanded with the heat of series of explosions, scattering billions of billions of centers of small fire balls of photogens which in the future will form the nucleus and axis of every material object.

Photogens stretched their wavelength and increased their population; space was inflated by the fire ball, which kept diminishing its temperature and pressure, transforming to lower energetic level of photogens progressively.

The photogens are the transformer of the Original Energy. It is energy-matter that could be transformed from energy to mass or from mass to energy. It could be energy-mater or wave-particle packet, the intermediate ingredient between energy and mass. It possesses potential space-time energy and kinetic time energy. Inside the core of the galaxy photogens will transformed to photons, subatomic particles, atoms, composites, deriving stars and all kind of heavenly object.

Photogens are the matrix of photons. They are billions of years in the core of the stars irradiating photons.

Photogens could act as energy or as messenger of the Original Energy to rule and put in order the universe. They are organized and form part of a system, the system that realizes the transformation of the Original Energy from an extremely compacted waves and enormous energetic frequency to the normal variety of waves that are common photons. So it is the transitional energy. This transitional period required a violent explosive inflation with a speed trillions of times faster than the speed of light, during the formation of the universe. If the Universe was really formed in seconds, that was at lease the speed required.

Space-time is an intrinsic inherent ingredient of the Original Energy and of the photogens. That is why with the explosive inflation, the Original Energy extending the wavelength by itself forming space-time which departs from the core of every nucleus.

The other characteristic of the photogen is: it constituted by extra high energetic waves. Depend on which phase it encounter, it could be detected or undetected. It could be extremely high energy or even low energy as common photon.

The most important and extraordinary characteristics of the photogens are: with the Original Energy they form the powerful nucleon of all existent material structure since microcosms up to macrocosms. They form the rotational axis of the atom, Earth, Sun, stars, galaxies and the universe.

Foremost, from the nucleus constituted by Energy O and photogen, every celestial body emanates radiations radial outward forming the energysphere hundreds times the diameter of the celestial body. It means, spacetime "curved" by the presence of a heavy celestial body in reality is the energysphere which confronts others energysphere forming the real limit and borderline between them.

That way photogens filled up all the disc of galaxies, also interplanetary, inter-spiral, intergalactic spaces, giving the spherical energysphere to all those structures.

Photogens also filled up the vacuum space and participate in all kind of energy-mass microcosm's activity; they could transform to electrons. Then, the positron appears converting both to the kind of photon that is necessary and gives back the energy to photogens continually; they also could become subatomic particles and convert to matter when the condition requires it.

But during the formation of the universe photogen played the most important function, photogens generated photons dividing them. That way photons filled up the entire universe during prolong period of time. Today that is still the way photogen produces photons from the core of the stars.

So photogen is a germinate energy from where the matter, the celestial bodies are formed. Even from the gas and dust of the nebula, miraculously almost from nothing, the Original Energy through photogen make the star appear because the photogens act on the microcosms.

Photogen is the master piece of the uncertainty; it is the key of the mystery of quantum gravity. For Quantum Mechanic vacuum is not vacuum there are virtual particles that fill up the space which in reality is the transformer photogen. But it is not science fiction, photon is built by kinetic and potential energy; one part wave other part particle; one half electron other half positron. Electron is instable, very sensitive to any change; any perturbation. Electron and positron could separate and travel toward opposite side as energy and disappear or join together and appear as photon and chains of photons.

The Original Energy theory states vacuum space is characterize by the mass emptiness where the kinetic time energy of the electromagnetic photon energy is the principal ingredient. While the Original Energy expands and accelerates the expansion of the universe by stretching the ultra energetic wavelength; photogens hold the material universe together transforming energy to matter continually, maintaining the order and equilibrium. That is why the universe gives the Steady State impression!

Photogen not only exist in the nucleus of the galaxy, it is spread and fills up all the interstitial spaces forming the invisible part of the galaxy, forming the energysphere of any heavenly body.

PHOTON

The Photongenesis or Photogenesis theory (The Original Energy theory) states: photon is the most fundamental constituent of the universe; photon is the most fundamental constituent, the cornerstone of everything that exist inside the universe; everything initial with photon and everything finish in photon. Energy is built by photons; mass is built by photons. That is why energy and mass could be interchangeable under certain condition of acceleration and time. Photons are built up with kinetic time energy and potential space time energy that is why photons have wave particle like duality; that is why photons intercede in every transformation, action or reaction of the universe. Photon not only is the inert object builder; photon is also the builder of every living being. Foremost, photon is an intelligent being that is why there are so many intelligent beings in the world!

Photons are formed with electric and magnet; by oscillation or rotation they form the electromagnetic field. They are the real force that transforms the universe.

Photon is made by electron and positron, because electron is negative and positron is positive they neutralize each other making photon to no charge. Photon could divide itself converting to free electron and positron and then electron and positron transform to less energetic photons again. Photon could reduce its size converting to a particle like or a point like. It is wave that can vibrate, oscillate, twist, spin, swirl, squeeze and stretch.

Photon is the quanta of the electromagnetism; combining with electron they are responsible of all activities in the universe. It is the carrier of the energy of all range wavelengths spectrum of the electromagnetic radiation from low frequency up to high frequency: radio wave that are used in radio and television stations; microwave that is used in oven for cooking; infrared that appear in incandescent light; visible light (including red, orange, yellow, green, blue and violet); ultraviolet radiation which cause burning of the skin; X rays that use in medicine; the last one the harmful gamma rays in atomic bomb and nuclear reaction. Higher its frequency more particle like its behavior; lower its frequency more apparent its waves behavior. Photons could transcend to higher frequency than gamma rays absorbing electrons becoming to photogen or E ray photons. Energized by more electrons they could transcend to even higher frequency becoming to ultra energetic photons which is the Original Energy.

The electromagnetic waves range from one kilometer or more for radio waves, nanometer for X ray and up to millionth nanometers for gamma ray. Extra and ultra gamma rays are even shorter. If we considered before the photons scattered at the initial stage of the universe they were compacted waves, they should be even shorter and their frequency trillions of time higher.

Photons transfer the energy to electrons becoming lower frequency and longer wavelength photons, this was the principal ways the ancient extreme high energetic photons becoming to today's photons.

Photons need a medium to propagate which is the electromagnetic structural mesh of the universe constituted by the Original Energy. Photons and electrons form partnership carrying and transforming energy, filled up the entire universe. It is measurable by its frequency, speed, wavelength and amplitude.

The electromagnetic has two components electric charge and magnet which create each other: by rotation the electric charge induces the formation of magnetic field; then the magnetic induces the formation of electric. They are perpendicular to each other in vacuum flying with the speed of light, creating the electromagnetic field. Any existence of the universe possesses these two components because all existence is built by photons, by electromagnetic field. The variety depends on the angulations of these two forces when photons penetrate different mediums.

Any range of photons travel with the speed of light in vacuum through the Original Energy structural mesh. But depending on which medium they transit, photon changes its behaviors presenting the characteristic of reflection, refraction, diffraction, interference and polarity by changing its frequency, wavelength and vector. Also depend on its size, photon has different actions: smaller the size, higher its penetration power.

Photon losses energy when it collides with high density atoms or particles that could absorb those energy.

Photons are emitted from the duplication of the photogen inside the core of the stars. Then they produce pair of electron and positron. From the nuclear reaction the subatomic particles, the hydrogen and helium atom are formed. Then the thermonuclear reaction of these elements enters in action, closing the photon, electron and positron production chain. It means, the radiation of subatomic particles, elements as hydrogen, helium are derived from even higher energetic photogens.

Photons have two components: kinetic time energy and potential time energy. Each could have the maximum value at the beginning or at the end of any event and demonstrate predominance of their momentum. For example: at the initial stage of the formation of a star, potential time energy was zero, it possessed only the maximum kinetic time energy; when a massive object is formed, before it decay, the potential space-time energy reach the maximum. The alternation of kinetic and potential energies is the real force of any activity and transformation of the entire universe.

The duality of the kinetic time energy and potential time energy of waves is the interaction and ratio between both: when the kinetic energy diminishes the potential energy increases; when potential energy diminishes the kinetic energy increases. This also reflexes the wave/particle duality of photon. The energy always conserves.

But while the kinetic time energy reaches its maximum in a close surrounding system of high temperature and pressure, the potential time energy would go down to zero. For exp, in a black hole, every existence is converted to photon; photons absorb all the electrons, transcend and become to photogens with extra high frequency. Everything become to kinetic energy, returning the energy to the universe.

During a long period of time, the new born universe was developed by compact waves which were extending; the kinetic time energy begun to transform to potential space time energy, inflating the space and forming the heavenly bodies. Since then and forever, the Original Energy forms the principal source of energy that expand the universe; the photogens form the principal source of energy-matter that transform to heavenly bodies keeping them attracted together and photons the most fundamental elements that constitute everything. All of them constitute the united electromagnetic field. Therefore, the electromagnetic force has been the most important force since the formation of the universe.

Photons are the elements that constitute the entire electromagnetic spectrum which travels with the speed of light in vacuum. Anything in the universe, since subatomic particle, electron, atom, planet, star, up to galaxy rotates and spins, incorporating to the electromagnetic field because the fundamental action of photon.

Photons have grand variety of characteristic nature: by stretching their wavelength they made possible the inflation, expansion of the universe and form the spectrum of wavelength from gamma rays to radio rays, constituting the electromagnetic field; the size of the photon is proportional to the wavelength, they could compress or squeeze to a point and act as particle penetrating any place; they could duplicate similar to cellular multiplication; inside the matter or heavenly body they attach one pole to the central point of the energy axis combining with electrons becoming to unidirectional photogravitons, constituting the gravitational force. They could change their speed when they pass through different medium but when they return to the vacuum they acquire the speed of light which reflex that photons possess intrinsic energy. Since the formation of the universe, photons have transferred their energy to electrons converting to less energetic photons. The process could be reverse lower energetic photons could become to higher energetic photons, even could transcend becoming to extra high and ultra high gamma rays. These are the most fundamental properties of photons and of the Original Energy Theory.

Photons might continue to stretch, a new variety of photon with longer wavelength and less energetic, lower frequency than radio photon or different variety of coherent photons could appear in the future. Biophoton could be an example.

ELECTRON

According to the Original Energy theory, Photongenesis is a mutual generation between photons and electrons, by division and duplication they generate each other and less energetic photons are formed. This phenomenon occurred during the formation of the universe and is still occurring inside the nucleus of all stars and galaxies with the nucleogenesis and nucleosynthesis procedure.

Electron is a negative charge elementary particle; it has no substructure or components. The spin, the angular momentum of the electron is a half integer value making it fermions. Electrons rotate around the nucleus of proton and neutron at different energy levels while they spin billions of times every second forming a cloud of energy.

When electrons vibrate they create magnetic field, when electrons move back and forth or vibrate their electric and magnetic field change forming the electromagnetic field.

In 1924 De Broglie hypothesized that electrons and all massive particles have wave particle duality as light, it means as photons. Most of the electrons were formed during the initial formation of the universe. Electrons always form partnership with photons in any physical, chemical, biological, thermal action, reaction, evolution, transformation since the earliest epoch. Photongenesis, Nucleogenesis, Photonsynthesis,

Nucleosynthesis, atom formation, black hole recycle system (Nucleolysis), all these are good examples.

The Original Energy theory has postulated that the energysphere that radial from the core of every star, every galaxy, every heavenly body is formed with the combination of photons and electrons. That way photons and electrons form the electromagnetic field, form the gravitational force and spacetime! To prove and demonstrate how they combine to make all these work is the highest challenge.

The electron is the smallest, lightest particle and as fundamental as the photon. During the initial Photongenesis process each photogen generates a pair of electron and positron which reacted becoming to a pair of less energetic photons. These new photons were converted to electron and positron, then they converted to a pair of less energetic photons; photons transformed to electrons and positrons again. These endless phenomenons occur inside the nucleus of stars and galaxies up to today. What implies the wavelength got longer and longer whiles the frequency become weaker and weaker.

The magic occurs right here in the transformation of photons which is energy, to electrons which is matter. The secret resides in the transformation of kinetic energy to potential energy and from potential energy to kinetic energy; from invisible energy to visible energy.

The entire secret of the origin of the universe; the origin of every existent inside the universe; the origin of life are enclosed in the transformation of the most tiny elements photons and electrons, from energy to matter and from matter to energy!

The entire universe was originated from the same ingredient which is photon. Consequently, everything obeys the same physics, chemicals, electric, magnetic, biologic laws. Everything is similar and homogeneous in the universe no matter whoever or wherever is observed; the temperature and the microwaves background radiations are isotropic because inside the universe everything come from the same origin which was the photon's fire ball.

Electrons are energy carrier, when they are accelerated they could absorb or release photons. Hence, electrons also could energize photons as a reverse procedure of Photongenesis which is the Photonsynthesis procedure. Inside the black holes all matter, ashes, radiation, light every existence are convert to photons and electrons a Nucleolysis phenomenon. Electrons joint to photons energizing them, converting them to each time higher frequency photons from radios, microwaves, up to gamma rays. In a further stage electrons could even energize photons until they become to extra or ultra gamma rays, the Photonsynthesis procedure.

Same phenomenon could occur if the universe reaches its end which would be the fate of the universe. All the material existence of the universe would be converted to photons; electrons would energize photons from radio ray to microwaves, then from microwaves to infrared, then from infrared to seven color of visible light. Then electrons keep energizing photons converting them to ultraviolet rays, then to X rays and gamma rays of the electromagnetic spectrum. Electrons could energize photons to even higher frequency to extra and ultra gamma rays which wavelength are over one ten millionth cm. Photons with lower frequency are also could convert to electrons and energize higher frequency photons, reducing photon´s population.

Electrons are the masterpieces that make possible the transformation of photons from extremely energetic to weak energetic photons or opposite from weak energetic photons becoming to extra and ultra energetic photons. These are the procedures that make the entire universe to transform.

SPACE TIME

At the beginning of our universe when time was cero, space was as small as the energy embryonic formation. Spacetime was inhering inside the Original Energy. The Original Energy comprises the maximum value of energy. The potential energy and the kinetic energy were in rest without vibration. They were stored with the maximum density as extremely compact wavelength photogens, without electromagnetic field.

Once the photons were free, they flied out, forming the electromagnetic field and converting to kinetic time energy, unfolding and extending their wavelength, filling up the entire new born universe. Photons traveled even faster than the actual speed of light, inflating, expanding and forming the space. Time was inhered in the kinetic energy; they are movement, rotation, action, evolution and transition.

There was no matter existed yet. The kinetic time energy began to transform to potential space-time energy where space is inhered; potential space-time energy becomes stronger and stronger while kinetic time energy looses strength. The potential energy formed the electromagnetic structural web of the universe.

Material space was formed while the potential energy grew through the transformer photogen; the volumetric, baryonic matters formed by potential space time energy occupy part of the space. Potential space time energy possesses dimensions: length, width, height, time and radius.

So space could be filled with matter, with potential energy or it could be filled with kinetic energy.

We can state that potential space time derived from the transformation of kinetic time energy during the formation of the heavenly bodies; matter, space, time derived from potential spacetime energy and both constituted electromagnetic field. It means matter and its volumetric spacetime is built by electric and magnetic ingredients which are photons. That is why energy and matter could be equivalent because they are built by the same ingredients which are photons. That is why energy and matter could be interchangeable because they both are magnet, electric, dimensional space and time, they could be quantized because they are photons.

According to the Original Energy theory any heavenly body possess energysphere emanated radial from the core of the stars or the nucleus of the galaxies; matter and energysphere (spacetime) form an inseparable unity just like body and soul form an inseparable unity. The Earth is formed with the mass and the energysphere emanated from its core; the Sun is formed with the gaseous mass and the heliosphere, which is the energysphere originated from the core that form the spacetime around of the Sun and so on. Together form the attractive gravitational force. It means, the gravitational force, not only depends on the amount of matter as Newton affirmed, neither depends on only the curvature of spacetime as Einstein established. Heavenly body and its energysphere together form a rotational unity.

Vacuum is not empty; it is constituted by the Original Energy mesh and the transformer photogenic energy which constitute the most important part of the universe. Space time is the place of action, interaction, confrontation and division between the heavenly bodies where the energysphere interlaced. That is why every heavenly body could be "suspended" and keep rotating in its orbit in the space.

Spacetime do not bends the mass less photon; light suffers deflection when light strikes in the energysphere of the Sun, black hole or galaxy that interposes between the far star light and the observer.

The energysphere (space-time) is periodic, dynamic; they are energy dependent, the energy that is contained in the nucleus of the heavenly body and emanates radial from the core making possible spin, rotation, evolution, motion, transformation, gravitation, decay of everything. Time is the period of action, movement, progression, termination. Time comprise the process of gestation, born, live, evolution, mature, dead and recycle. When energy exhausts, time finishes and vanish, leaving ashes and death until it is recycle; when energy exhaust the four dimensions volumetric space collapse and its space time is meaningless. It means when the concentrated Original Energy of the core is exhaust all matter and energy had already converted to low frequency energy of photons and electrons containing inside the energysphere.

According to Photogenesis theory Energysphere departs from the core of the heavenly body speeds out through it and forms spacetime. It is inherent in every activity, every event: thermal, chemical, magnetic, electric, mechanical and biologic because energysphere which includes spacetime is potential or kinetic energy transformation.

Energysphere (spacetime) is relative to every heavenly object, objects spin and rotate in a determinate space and in a specific limited time; objects change, evolve, move, and decay constantly; they have momentum under a constant transformation. Hence, space-time depends on the heavenly body's change and move constantly too which is a specific orbit. There is no a fix spacetime because everything rotates, moves and changes their position.

If we considered the cyclic transformation of the universe, space-time could contract or inflate periodically with the Original Energy together. We got to be conscious space-time of our Universe is finite. But outside of our Universe, space is infinite and uncountable universes could exist lots of times.

According to the Original Energy theory energysphere (spacetime) is always curve not just in the presence of the heavenly bodies cause by their weight. The curvature is because the universe itself and every object spin, revolve and because the universal sphericity. Depend on the energy the heavenly object possesses, it possesses specific orbit. Hence, energysphere (spacetime) is orbit which determines locality, position, curvature and radius. Orbit implies gravity, electromagnetic attractive energy.

Foremost, from the nucleus constituted by Energy O and photogen, every celestial body emanates radiations outward forming the energysphere tens or hundreds times the diameter of the celestial body. It means the axis of the Original Energy form the energysphere and rotates the heavenly body together; heavenly body and energysphere are not independent entities. The celestial body does not rotate through the "curvature" of spacetime because the heavy weight wraps it. The reality is the rotating, radial energysphere comes from the nucleus of every heavenly body, forms spacetime, confronts others energysphere of others heavenly bodies, forming the real limit and borderline between them maintaining them in a specific position and orbit.

That way, all the heavenly bodies and their energysphere are "touching" each other, maintaining "suspended" in the space, living in a concordance vicinity.

TEMPERATURE

The Original Energy Theory states that before the initial of the new universe the temperature was close to absolute zero, close to 0 Kelvin. That is why the energy could be compressed; that is why the contraction up to singularity was possible. At cero K, photons are in rest, with cero speed and acceleration, they could not move; their electromagnetic field was extremely reduced, photons becoming a point and keep in hibernation.

It is inconceivable under extremely high temperature, trillion of trillions degree, matter, in any state: solid, gas, liquid or plasma could be compressed infinitely up to a point! Just because the volume of the matter in any state under the extremely high temperature would increase not reduces up to an atom! The volume of our universe should be billions of times bigger, inflated only with heat which is kinetic energy; an exactly opposite process to singularity! If the universe derived from a post singularity stage, it should be developed from close to 0 Kelvin.

According to the Original Energy Theory during the embryonic period, photons were contained close to cero K, when the duplication commenced, photogens vibrated, warming up, elevating the temperature. The compressed, cool energy was inflated because electron and positron appeared and underwent to nuclear reaction. Hence, kinetic energy reached its maximum, elevating extreme high temperature that reach trillions of trillions of times of the Sun temperature, igniting a huge explosion.

During the formation of the universe, temperature was the most outstanding factor for any state of the matter. After electrons and positrons annihilation episodes, extremely high temperature did not permit the formation of matter. That temperature could melts, inflates, burns up and evaporates everything, converting it to heat immediately after its formation, only kinetic energy would persist which were photons. On the other hand, under extremely high temperature and violent vibration, gluon cannot exist, cannot work sticking subatomic particles, binding proton with electron, matter could not form. That is why at the initial formation of the universe, the universe was only a fire ball, a ball of kinetic energy.

At any fundamental transformation, temperature is elevated and explosions could appear. That was the fundamental factor that causes the violent explosions and inflation during the fast formation of the universe, the real moment of Big Bang episode. This is the fundamental factor that determined the only existence of the electromagnetic force stem. This is the fundamental factor that determined the inexistence of the gravitational force which derived from electromagnetic force posterior to the formation of subatomic particle atoms and mass!

Where did such heat go away? And how the heat could be transmitted without any particles or fluid, in an absolute empty universe where only high energetic photons radiation existed?

There are three ways in which thermal energy can be transferred:

By conduction, in which particles of a medium vibrate and transfer the heat throughout the particles of the object. For instant, in a close water heater system, the water is heated in one side; the heat is transmitted through all the system. But the water particles are not running. There was no gas, no liquid no material particles conductor during the universe formation period, so it could not be by conduction.

By convection, where the thermal energy is transfer by movement of matter particles, it used to be particles of gas or liquid. For example: the cold currents of air that come from the North Pole, the cold temperature is transmitted directly by the air particles. Because the same reason, there was no any material substance fluid inside the new formed universe, it could not be by convection.

The third possibility, the Photogenesis theory affirms that the heat that existed inside the fire ball was transmitted by radiation and this is the only way possible the heat was transmitted during the period of the formation of the universe. In vacuum, the thermal radiation energy is transfer by itself. There is no substance or current of substance needed. The energy is transfer from one place to other place of the space through the oscillation of the electromagnetic fields. Photons transport energy from one point to other and this was the unique way possible that the heat of the new born universe was transmitted! Inside of the fire ball of the new born universe only heat and light existed which were photons. The unique existence of the cosmic background radiation from the initial time confirmed this statement!

In that stage photons were distributed isotropic and homogeneously inside the fire ball, the new born universe was a closed, dense and extremely hot sphere. After series of explosions, billions of small fire balls were spread out altering the homogeneity and the temperature becoming anisotropic permitting the formation of subatomic particles.

In any black body the entropy processes appear cooling the extremely hot universe down. On the other hand, the violent expansion caused the temperature of the universe diminished progressively by spacing. The decreased temperature and the increased inflated space determine the extension of the wavelength of photons which imply diminish frequency of the photons. That was the force that inflated the universe at the same time.

Until today vacuum space is the best insulation medium, because there are no matter, no fluid, with a temperature of cold 2.725K everywhere all the time, which is the remnant cosmic, microwaves radiation of the initial explosion. The reason that the temperature of the universe is still isotropic and constant is because the heat of the stars, supernovas, galaxies, black holes and their explosions are kept inside the energysphere by the vacuum insulation.

Vacuum is very bad heat transmitter. The heavenly body's temperature is kept isolated. For instant: the Sun's heliosphere is insulated; its temperature never escapes by convection or conduction to the vacuum interstellar space; neither by radiation because the Sun radiation is stopped at heliopause by interstellar radiations.

Temperature and pressure is a determinant factor in every step of the evolution, transformation of the energy or the formation of the universe. As we know, at the initial extremely hot stage, only energy could exist. Outside the hot sphere there was no pressure, so the new born universe expanded and inflated freely. The universe tended to be cooler because any transformation of energy loss heat and because the density of the energy diminished by spacing. But it was still hot enough inside the nucleus to permit the chemical and thermonuclear reactions to form atoms, composites and mass, In accordance with the expansion and descends of temperature; the universe was inhabited with matter. Most of the elements were formed during that period, including the heaviest elements.

Temperature is kinetic energy; the cosmic background radiation is kinetic energy. If since the formation of the universe the frequency of the energy has been diminished, it is logical that temperature has been changing decreasing also which had been a lot higher than 3K. But in vacuum the temperature is still isotropic because the heat of the stars, galaxies, black holes and all the successive explosions are isolated by the vacuum while the vacuum temperature has kept invariable.

The temperature of the universe is isotropic in all directions; it means the temperature of the universe is isotropic in all 360 degree spherically and in large scale the cosmic microwaves background radiation is distributed homogeneously. The photons of the cosmic background were radiated straightforward, the opposite direction radiations have the same temperature because they were originated and emitted from the same small fire ball, the same background emission, had the same temperature initially and because the effect of revolving force mixes and homogenizes.

Any way the high temperature was the factor that contributes the agglomeration of subatomic particles to form all kind of chemical elements, to form celestial bodies.

The existence of matter with higher temperature makes the space clumpy. Inside of the core of any star or galaxy or black hole where the thermonuclear reaction takes place, where the photogen energy resides, the temperature is still extremely high.

The isotropic temperature implies that the revolving Original Energy gigantic labor which regulates and maintains the temperature of the universe. The continue explosions of the extra energetic gamma bursts, black holes and supernova not only contribute to recover energy to the universe with their radiations but also reply heat to maintain the constant temperature. The temperature of the photons from those sources, from any star is kept invariable through the vacuum while there are no any medium to interfere with them.

Temperature is the fundamental factor in the formation of life on Earth. Earth is localized in the comfort zone of the solar system with optimal and adequate temperature for the development of life. The variation between day and night temperature and different stations is not as extreme as in other planets. Even on our Moon, day and night temperature is extremely different that not permit life development.

Earth is localized at the comfort zone of the galaxy and the universe; we are not localized close to the nucleus of the galaxy neither the center of the universe were temperature, vibration, turbulence are extremely high.

It is believed that the heavy metals are formed only in supernova; supernovas are formed at the end of the recycle chain, and only one per cent of the death stars are converted to supernova. According to the Original Energy theory most of the heavy elements were formed at the beginning stage of the universe, the hottest epoch of the universe. Most of the heavy elements are formed by nucleogenesis nucleosynthesis and localized in the nucleus of all heavenly bodies. Posterior to that time, it is more likely that the heavy metals formation depends on the temperatures that the core of the stars and galaxies possesses; it means depend on the energy they possess. Higher the core temperature more heavy elements could form and possesses; then the heavy elements decay undergo to fission releasing energy, elevating even higher temperature. It implies also the heavenly body formation, higher the temperature bigger and more solid the heavenly bodies could formed; more heavy metals as iron, diamond, uranium they possess in their core.

Posterior, supernovas contribute to the formation of heaviest elements localized on outer layers of the heavenly bodies and space.

MATTER

Matter is a compressed expression of energy; it comprises three levels of Original Energy: ultra energetic photons or O photons, extra energetic photons or E photons and the common electromagnetic spectrum photons or C photons.

The universe begun as an embryonic energy formation, it was converted to a fireball, during that Photongenesis epoch there was no matter but heat of photons. It took hundreds of thousands, maybe millions of years for the photogens and photons to be distributed homogeneously in the entire universe while the dense and compact waves become fragmented and stretcher waves. Outside of the fire sphere there was no pressure, there was no temperature, the space inflated and expanded freely without any resistance.

After series of explosions, the big fireball converted to billions of small fireballs and was disseminated. The small fireballs alter the isotropy and homogeneity that originally existed, producing fluctuations on the concentration and temperature of energy. Multiples explosions disseminated billions of revolving energy, which comprised primordially photogens. They were spread progressively and converted into the nucleus of embryonic stars and galaxies which constituted the Nucleogenesis procedure and that was the Nucleogenesis phase.

The fire balls left abundant O photons, which were ultra energetic gamma rays; they released electrons and positrons producing extra energetic gamma rays. Pairs of photogens were produced abundantly with less and less frequency each time and then become to pairs of new photons, saturating the universe. The extra energetic gamma rays, the photogens are built by magnet and electric energy, by kinetic and potential energy. While the temperature decreased the kinetic energy diminished. The wavelength of the photons got longer and longer, while the universe expanded, inflated and got cooler, until the formation of wavelength close to photons of the actual electromagnetic spectrum.

From the spinning nucleon, the plasma of photogens was transformed to plasma of quark and gluon and others subatomic particle; subatomic particles are spread out and getting cooler, beginning the combination of quarks and leptons, forming protons and neutrons. The ionized state changed permitting protons and neutron to bind with electrons. The atom and mass begun to formed from the most elementary particles, ironically the no weight and no mass photons.

The nucleus of embryonic stars and galaxies undergo to nuclear reactions developing extremely high temperatures and pressure again. Nucleosynthesis procedure took place converting the light chemical elements to heavier of all kinds of elements. That was the Nucleosynthesis phase.

In that extremely high temperature stage, light chemical element atoms overcome to fusion process becoming to heavier and heavier chemical elements. Under the centripetal rotation of the nucleus of the Original Energy, the heaviest elements are concentrated in the core around the nucleons of the Original Energy. The lighter elements are spread out forming the outer layers of matter and atmosphere of heavenly bodies.

Matter or mass is potential space-time energy; it is tangible object, can be measured because it occupies space and possesses dimensions high, width, long, radius and time.

For maintaining a functional structure it consumes energy. Therefore matter that comprises space, time and potential energy decays transforming the potential time energy to kinetic time energy again, energy that incorporated to the energysphere. These constitute the Nuclear Reaction phase.

Matter has three ways to transform from potential time energy to photons kinetic time energy:

1). by radioactive decay: a radioactive atom as uranium emits spontaneously a particle which transforms to an atom and release energy as photons;
2). by nuclear fission: when a single massive atom splits out into two smaller atoms and releases energy which is photons;
3). when two smaller particles bound together and form one bigger particle and release energy. This is the thermonuclear fusion process releasing photons.

Most of these happen successively in the nucleus of the stars. These alternations between potential and kinetic energies confirm that mass is built up by photons and their electromagnetic field. The alternation from kinetic photons to potential photons or vice versa constitute the continue transformation of the universe.

During these Photongenesis, Nucleogenesis Nucleosynthesis and Nuclear Reaction procedures, the new photons and electrons were formed each time weaker and weaker. The antimatters that were formed are each time weaker and weaker also. That is the reason why the matter overcomes to the antimatter; united with the outward expansive and inflation force, the universe could form expanding against to any contractive introversion force. It means the electromagnetic outward force overcome to the gravitational contractive force.

These constituted the initial formation of matter, from photons and electrons to subatomic particle. Then from subatomic particles to atoms, then atoms to atoms compound, mass, moons, planets, meteorites, asteroids, stars, solar systems, galaxies, cluster of galaxies up to the complete material universe.

In the macrocosms, matter is characterized by the gravitational force which is attractive to the geometric center point. As it has been pointed out before, matter is built up by potential time energy which is built up by photons electrons partnership, combining they form the electromagnetic field which is an outward force. We could affirm that matter and its gravitational force in reality are built by the combination partnership of photons with electrons too.

Photogens not only exist in the nucleus of the galaxies, they are spread out and fill the interstitial space, the invisible part of the galaxy, from where they transform to matter or energy.

Our solar system was formed from the Original Energy from the rotational nucleus of the Milky Way galaxy. Under the extreme high temperature and pressure, the energy was transformed to waves of photons, electrons and subatomic particles which coalesce generating the primeval plasma. Part of the plasma suffered differentiation, transforming to the hot red nucleus with spinning rings of matter around.

The nucleus transformed to the Sun and the rings continuing to coalesce, while they were spread out they got cooler transforming to planets.

The mass that could not convert to planet were conglomerated around the planet, forming rings which were conglomerated to the hotter groups, forming the moons. Our Moon is not an exception.

Because matter tend to decay, the potential time energy that posses would be spent and the heavenly body would out of fuel. The star or the galaxy would be recycled by black hole converting to kinetic energy. This is the Nucleolísis phase.

In the phase transformation, explosion could happen become to a supernova. New energy, light elements, heavy elements are formed again.

Because the constant expansion, the renovation of energy by black holes and reposition of matter with supernovas, during long period and up to today, all matter, all heavenly bodies, energysphere and energy keep a harmony order with homogeneity and isotropy distribution. The entire universe enters to an apparent Relative Steady State phase.

The Original Energy theory states the energy of photogens keeps transforming to matter which is the power that keep the heavenly body attracted together. At the same time, the Original Energy keep forming less energetic photons, their wavelengths keep extending; the black holes keep recycling matter to energy. Hence, the universe would keep expanding.

Consequently, the Original Energy through photogens forms continually matter which is the force that holds all the heavenly bodies of the universe together. This is the factor that made the existence of the doubtful dark matter unnecessary to avoid the disintegration of the material universe. On the other hand, by extending the wavelengths as it has been happening since the formation of the universe, the Original Energy expands the universe making the existence of the doubtful, hypothetic Black Energy unnecessary.

PHOTOGRAVITON AND PHOTOGRAVITY

As it has been pointed out before: according to the Cosmological Model, every event, every observation, every conclusion has been attributed to the gravitational force, even nobody knows exactly what forms gravitational force, what is graviton? Graviton has not been discovered. Einstein´s General Theory of Relativity established gravitational force is an illusion; there is no force at all! It affirmed the curvature of the spacetime is what makes everything rotate, is what makes everything attract together. Gravitational force is inertia and General Theory of Relativity is the cornerstone, the origin of the Cosmological Model.

Nevertheless gravitational force is the foundation of the modern scientific astrophysics and astronomy. The gravitational interaction has been considered as a force. Since the late 16th and early 17th centuries Galileo Galilei showed that gravitational force accelerates everything with the same rate; is the air resistance that makes the difference between the velocities of falling objects. Newton in 1687 published Principia establishing three monumental physic laws, confirming that gravity is a force which is directly proportional to the product of masses and inversely proportional to the square of distance. Newton considered the gravitational force resides inside the central geographic point of the mass, forming a vector. Newton made almost perfect mathematic calculations that are still useful in the modern astronomy.

The Photongenesis theory established that photogens, photons and electrons are the most fundamental elements that constitute everything in the universe. Gravitational force is not the exception; it results from the partnership combination and mutual formation between photons and electrons.

During the universe newborn epoch only ultra high and extra high frequency photons beside electrons existed; the electromagnetic force was the unique existent force. After the final photon fire ball dominated epoch, subatomic particles arose from the transformer photogens, constituting the nuclei of quack and gluon, light elements begun to bind. After that, the gravitational force was differentiated from electromagnetic stem force but it was extremely weak.

All the atom and object of the universe are dipole, under normal condition and temperature they form a magnet bar; because matter or energy derived from the same ingredients (photons and electrons), gravitational force is a kind of electromagnetic negative attractive force, constituted by chains of bidirectional photons combining with electrons.

The most evident that the gravitational force is formed by the combination of photon and electron is in the core of the stars. The nucleon of a star is formed by Original Energy; photogens generate photons and electrons, through them they form all kind of subatomic particles. Then proton and neutron bind electrons forming atoms since light chemical elements up to heavy elements.

Under the high temperature, heavy atoms as iron, diamond, uranium are formed and surround the nucleon forming the positive charge core of the star. The light elements as hydrogen, helium up to carbon undergo to thermonuclear fusion producing chains of pairs of less energetic photons and electrons.

On the other hand heavy elements undergo nuclear decay or fission releasing photons and electrons. This way, kinetic energy transforms to potential energy and potential energy transforms to kinetic energy. During billions of years these processes are repeated again and again releasing photons and electrons. That is why photons and electrons could delay billions of years inside the core of the star or galaxy until they escape with outward pressure.

Photons and electrons are polarized, one pole is connected to the central geographic positive point of the Original Energy of heavenly body; the other pole forms the unidirectional negative radial chains from the core of the star constituted by electron-photon-electron with spin -2.

This is the negative attractive radial formation force from the nucleon of mass radial to the space around of the planet, star or galaxy.

Hence, in reality gravitation is made up with waves of dipole photon and electrons chains. Shall we assign a name as photograviton for the gravity messenger the graviton and photogravity as gravitational force?

The Original Energy states that the universe is derived from energy not from mass. Hence, there were no subatomic particles but photons and electrons; even less possible molecules or mass existed at the initial stage of the universe. Subatomic particles, atom or mass were formed posterior to the inflation of the universe, after the nucleogenesis and nucleosynthesis are well established. The extremely high temperature begun to descend but not too much making possible the formation of elements from light ones up to heavy ones; high enough temperature made possible the formation of heavy elements.

Once molecules and matter were formed and the temperature descended, the gravitational interaction appeared. Matter coalesced to form heavenly bodies.

We can deduce the attractive Graviton is made by the combination between dipole photon and electrons. That is why graviton as elemental particle has not been found.

According to Newtonian theory gravitational force depends on exclusively on the mass between two objects which is directly proportional to the product of mass and inversely proportional to the square of the distance; it is only an attractive force. It has been believed widely that because the gravitational attraction, the moons rotate around the planets, planets rotate around the stars, stars around the nucleus of the galaxy, galaxies rotate around the cluster and the clusters rotate around the super cluster. Even convection phenomenon, tides of the sea are believed cause by the gravitational effect. But the spherical symmetric heavenly body should be considered as all the mass are concentrated to a point inside the geographic center, forming an attractive vector to reach any other central point of other objects.

Meanwhile Einstein's General Relativity considered that the Newtonian gravitational force is an illusion and the gravitational force results from the curvature of space time. The mass cause the space time to warp and the heavenly body rotates through lower gradient level by inertia without any real attraction force.

The Photogenesis theory states: any mass since the atom, star, galaxy up to cluster of galaxies, any heavenly body and their energysphere radial outward the heavenly bodies form an inseparable unity. Mass is built by magnetic and electric ingredients which is potential time energy. Meanwhile, the energysphere is kinetic time energy. Together matter and energysphere form the negative attractive energy field emanated from the mass geographic center which constitutes the real gravitational force.

We got to point out, the energysphere initial from the nucleon of the heavenly body as star or galaxy, it is not just a donut from outside of the heavenly body as spacetime or as a carpet over where the heavenly body rotates; it extends far away from the mass of solar system or galaxy. Just as body and soul of human kind, heavenly body and its energysphere form an integrated unity.

Hence, the Original Energy theory considers: By any mean they, mass or spacetime individually could act as the gravitational force which straightforward as a vector attracts any object on its reach to the nucleus. By the same reason, any heavenly body is formed with massive object and the energysphere that radial from the core enclosing it around. Earth possesses the geoenergysphere; the Sun has its helioenergysphere; Milky Way Galaxy has its energysphere. The energysphere could be tens or hundreds of times bigger than the material part, depends on the energy it possesses inside the core. Consequently any heavenly body, the solar system, the galaxy and the universe are bigger than the size that has been calculated or look like. The diameter of the universe could be more than 300 billion light years in all directions because the existence of the universe energysphere.

PHOTO GRAVITATION

Mass is derived from energy, it is made up with potential space time energy. It means mass derives from electric, magnetic and time. Therefore it is constituted by electromagnetic field and its quantum is the photon combine with pair of electrons constituting the dipole gravitational force. Gravity is an attractive force field that acts in distance; its vector is directed toward the center geographic of each object or heavenly body, but the electron-photon-electron chains are expelled outward. Hence, there is no wave or particle of gravity, the graviton could be found inward from one to other object!

For instant, the axis of the stars as our Sun is a rotational charge bar of the Original Energy generating electric field which electrifies everything in its reach; the axis rotates continually inducing the formation of magnetic potential which magnetize everything inside the solar system. Then the electromagnetic field is formed.

The axis is constituted with energy O which releases photogens progressively. From photogens derive electrons, and positrons; they undergo to nuclear reaction, elevating the temperature, forming chemical elements from light to heavy ones progressively. The axis and the core become to positive charge. Heavy elements undergo to decay, fusion take place delivering radiations which are photons.

The nucleuses of stars develop extremely high temperature and pressure. At the exact geographic central point of the stars, the Original Energy generates photogens and the photogens generates photons; photons generate pair of electrons, then electrons generate pair of photons endlessly from extremely energetic photons to less energetic photons during billions of years.

Photons are no charge but electrons possess negative charge; photons with pair electrons are polarized and align one pole to the central geographic point of the star, forming electron-photon-electrons unidirectional negative attractive radial chains.

These radial negative polarization formations of photons and electrons form also the electric field which coincide with the gravitational field and obey the inverse square law in distance. Hence, in reality gravitation is made up with the alternative waves of electrons-photons-electrons which form the negative attractive force. Shall we assign a name as photograviton for the graviton formed by photon and pair of electrons whiles their interaction as photogravitation force?

The generation of photons from extreme high energy photogen to common photons from the core of the star is through the interaction with electrons which form an outward electromagnetic pressure. So, we only could detect photons and electrons emitted from the star or galaxy but not the messenger graviton of gravitational interaction.

Every electron in the universe always and forever spins at one fixed and never changing negative rate. The spin of the electrons is an intrinsic property; not a transitional state, it does not depend on the energy supply from outside neither the curvature of the spacetime around or under it. On the other hand, photons absorb or release energy combining with electrons forming inseparable partnership in all kinds of action, reactions and transformation of the universe.

All of the matter particles have spin equal to that of the electron. It means all matter particles have spin -1/2. Meanwhile the non gravitational force carriers as photons, weak gauge bosons and gluons possess an intrinsic spinning characteristic of spin -1. The hypothesized graviton should have spin-2. Hence, the photograviton formed by photon, which generates pair of electrons, should have spin -2 which is -1/2, -1, -1/2. That way photon generates pair of electrons, electron generates pair of photons, electron generates pair of photons and so on. They could form radial chains as long as possible according to the amount of energy that each heavenly body or any object possesses in the nucleus, which reflex the amount of mass and energy. They obey the inverse square law of the distance because the same amounts of photogravitons need to stretch their wavelengths to reach each other. It means each heavenly body connects their electric, magnetic photograviton chains up to their limit, forming the attractive force.

If this statement of the Photongenesis theory is correct the gravitational force is not unlimited as always has been believed, it depends on and limits by the amount of energy and mass that every heavenly body possesses. It demonstrates clearly that the gravitational force is electric, magnetic attraction force!

In the case of our Sun, the action of the unidirectional photogravitons reaches beyond the planets, moons, asteroids and meteoroids, beyond all the heavenly bodies and object of the solar system. The electromagnetic field reaches even farther; the electromagnetic waves confront the interstellar and galactic radiation forming the heliopause and reach up to the bow front.

Coincidently, because the Sun is potent electric charges, its electric field reach several times beyond the material objects, forming a matter-space-matter-space unit where every material objects are charged and aligned with the electric field of the Sun; it means, each planet and its energysphere make matter space, positive-negative unit. Same phenomenon occur with the magnetic field of the Sun, it magnetized all objects of the solar system. Hence, all of the massive objects align with the Sun´s magnetic field. Both, electric and magnet are dipole; one pole is bonded in the central point of the axis but the other pole form unidirectional attractive force from the core of the star. All of the planets have opposite charge and opposite pole that are attracted by the photogravitons of the Sun.

Contrary, because the planets are not strong enough they do not electrify or magnetize others planets but only their moons; planets are not align between them but only with the Sun, turning their out crust positive. It means the Sun electric and magnetic field has an exact perpendicular angle that attracts all the planets and objects with the negative attractive electric force inside the solar system. On the other hand, planets, moon and other objects have different orientation between them, or same electric charge and same magnetic poles, they are repulsive between them. The electromagnetic forces always are stronger at the equator of the Sun that is why all the planets are attracted at the plan of the equator.

Same phenomenon occurs inside of each planet. The central Original Energy electrifies and magnetizes everything and attracts everything with the radial chain of attractive photogravitons. The gravity force of the Earth reaches beyond the exosphere. Because the closeness of the Earth and because the opposite charge between the Earth and the Moon, their attractive force overcomes.

So, gravity (photogravitation force) of the Earth or gravitational force of any material object, are intrinsic forces, constituted by photogravitons of the Energy O.

On the other hand, the electromagnetic field, the mesh of the universe which inherent space-time, serves as the medium where the electric and magnetic waves travel. But it is not spacetime curvature "the force" that makes the planets, moons to rotate around the Sun; it is the conjugation of all solar system intrinsic photogravitation forces that make a functional interior unity, equilibrium and an isle defensive barrier to the exterior.

The gravitational waves are mechanical waves; they are formed by the rotation and displacement of the heavenly bodies through the space time fabric which is the electromagnetic structural mesh. It is the heavenly body with its intrinsic energy that spins, moving away producing waves like at the structural mesh. That is why there is no graviton or any messenger particle could be found. Its speed obeys the inverse square law of the distance.

The Original Energy gave origin to the four forces that exist in the universe: the electroweak force and the strong force were formed when the subatomic particles and the atoms were formed, acting in the microcosm level.

The gravitational force arose when the mass and celestial bodies appeared, which imply that the gravitational force is a secondary force that differentiated from the electromagnetic force, acting in the macrocosms level. All of them act according to the regulation of the structural electromagnetic network of the universe. The spacing between the heavenly bodies are ruled by the Original Energy, each one are connected and limited with the energysphere, which imply the energysphere is the real electric and magnetic forces that connect and limit all the material universe. All the planets are connected and limited by their energysphere; all the solar systems are connected and limited with their energysphere; all the galaxies are connected and limited with their energysphere and so on. That is why the universe is so ordered and all the heavenly bodies could be "suspended" in vacuum.

ROTATION AND SPHERICITY

The Original Energy is a revolving energy; it forms the axis and the nucleon of every existent energy formation, material object and of the universe itself. The axis is a magnet with opposite poles. The rotational motion of the magnetic force axis generates the electric force and the magnetic potential induces the electric potential, both rotate perpendicularly to each other. Magnetic potential and electric potential are equivalent. Hence, they make a perfect spheroid with a radial symmetric distribution of magnetic and electric charge, producing a tridimensional spherical electromagnetic field.

During the formation period, under the extreme high temperature and pressure, the entire universe was a fire sphere of photons, from where everything in the universe was originated in the radial form. Consequently, the universe was homogeneously spherical and everything has the same origin from the Universe Original Energysphere. Hence every formation has an electric, magnetic axis and an electromagnetic field that form the energysphere.

Electron as photon always spin with intrinsic energy forever, they are the most basic constituents of everything in the universe; they are the intrinsic energy that make everything rotates.

The atom is formed by a nucleon of proton, neutron, and a cloud of energy of electrons; they always spin making a perfect sphere. The heavy nucleus of proton and neutron is very small occupying an insignificant space; light electrons occupy an even smaller volume. What makes the atom looks like a big object is the spherical space forming by the cloud of energy levels. Moreover, the material that constitutes the atom might be 4% of the entire atom; the rest of the atom is the intrinsic Original Energy. That is the reason the matter of the entire universe only constitute 4%, the rest is energy!

The crust of the Earth is massive but the mantle and the outer core are liquid, because the rotation generates centripetal acceleration making the Earth a little bit ellipsoidal. But with the atmosphere and energysphere (electromagnetic field, and photogravitation field) they form a perfect sphere. The Sun heliosphere is an almost spherical bubble.

All the celestial bodies tended to adopt the spherical shape, caused by the revolving action of the Original Energy. The Original Energy rotates, mixes, compresses the mass and energy during the nucleosynthesis and during the formation of any heavenly body. That is the way it determines the density, the dimension, the composition, the aspect, the homogeneity, the sphericity and even the isotropy temperature of the matter. That is why the radius of the revolving dimension should be considered as the fifth dimension of the matter, as the others four dimensions: length, width, height and time.

Matter we can see in the entire universe appears to be concentrated in spherical shells, or bubbles. Every celestial body has mass, atmosphere and energysphere. The energysphere is an intrinsic kinetic energy, formed by photogravitation and electromagnetic field. Inside the sphere contain primordially photon radiation. In the case of our Earth, the Earth does not emit hydrogen or helium. The hydrogen and helium of the solar wind emitted by the Sun each time scarce up to vacuum, so their gravitational force is countless. The real force that encloses the Sun, the Earth and the solar system is the electromagnetic force and its field.

The sphericity is reflected as well in the spherical form of space time. Day time and night time varies because the tilt of the Earth but one day consists with 24 hours, no matter where you are located, at the equator or North Pole. Even the circle you made is a lot smaller at the North Pole; anyway it takes 24 hours a day to complete a day. This is a fact of photogravitation because the radiuses from the center geographic to the surface of the North Pole or to the equator are almost equivalent. Hence the rotational photogravities forces are close to the same, making time spherical, the same twenty four hours all around the globe.

Any heavenly body consist with mass and the energysphere, they form a sphere. The energysphere is emitted radial from the core.

The universe is homogeneous and isotropic physical and chemically, the space looks like the same everywhere, anytime; hydrogen, helium and others chemical elements are the same around the entire universe. The temperature is isotropic also. That is because the constant rotational action and the universal sphericity.

Because the electromagnetic rotational force of the Original Energy makes every existence tend to be sphere, the universe is an open sphere, it is not flat. Every celestial body rotates: the Moon around the Earth; stars around the nucleon of the galaxy; even the entire universe rotates around the axis of the Original Energy.

Light in reality dos not travel straightforward. Farther the source of light less notable its curvature, more likely it is straight since the radius is larger. For the same reason, the immensity of the universe gives us the impression that the universe is flat.

The sphericity is given secondarily by the gravidity effect of the gravitational force to each one or each group of celestial body. The gravitational force tends to contract the matter at all three hundred and sixty radial degrees but it does not rotate; it means the gravitational force is a contributor factor, not a determinant factor to make the rotating celestial bodies spherical or elliptical.

It says: atom, Earth, Sun, stars, galaxy and the universe are not spherical, they are flat. But they are spherical at least elliptical. That is because Energy O and photogen energy have not been taking in consideration. The atmosphere and energysphere should be taking in consideration; heavenly bodies revolve, revolving electric and magnetic field tends to be sphere.

Earth is a magnet sphere; the Sun is a magnet sphere; a galaxy could be magnet sphere, so on the universe is magnet sphere. Consequently, the electromagnetic field of all of them is sphere.

In the solar system most of the planet are situated around the equator of the Sun. same phenomenon we observe on the moons and asteroids, they are located around the equator of the planets. But with the energy around the solar system, it makes the solar system spherical, not flat.

The temperature of the universe is isotropic in all directions, it means in all 360 degree spherically and the cosmic microwaves background radiation is distributed homogeneously in large scale. It reflexes the sphericity of the universe.

The spherisity is universal. The perfectly spherical heliosphere is an example. It also demonstrates the size of the solar system is far away bigger, beyond Pluto, beyond material formation of the solar system. For the same reason Milky Way Galaxy is bigger than where the last star arrives. Hence, the observable universe is not the size of our universe; it is just the limit where we are capable to observe. Our universe reaches far away beyond the material limit; beyond the oldest galaxies and stars. It could be several times bigger than the observable universe extending with its energy. Just like the solar system. That is why the oldest galaxies are inside of our universe because the universe is older and bigger.

The gravitational lensing is a deflection of the light that comes from distant star. If we take in consideration the atmosphere, heliosphere, it means the energysphere of the Sun, these components act as a convex lens, deflecting the far star light and cosmic radiation. We can see the foreground image before we see the real star which is still in the background behind the Sun. The energy around the Sun makes a convex lens which causes a virtual image formation of the far star, galaxy, black hole or supernova close to the Sun, no matter how far away they are. Depending on the size of the background star and the alignment of the far star, the Sun and the observer, we could see two or multiple image of the star. On the other hand, light travel through the electromagnetic mesh which is slightly curve. Consequently, the gravitational lensing reflects the sphericity phenomenon.

The very massive supernovas have put in evidence by showing a vast zone of energy as glass all around the supernova. The Photogenesis theory has postulated that all the stars, galaxies, black holes, supernovas or cluster possess energysphere, convex lens like zone, responsible for the lensing effect.

The gravitational force is exclusively attractive force all around spherically radial to the Sun. If the gravitational force is responsible of the lensing phenomenon, light should be pull inward then leave, describing a concave curvature, but it is not. Photons that pass through the convex lens like energysphere of the Sun descript a geodesic curvature. This behavior confirms the existence of the energysphere around the Sun; confirms the spherisity of the Sun. It also put in evidence that it is a refraction phenomenon of the light entering to different medium from vacuum to the convex lens like different energy layers of the Sun, then leaves; there is no any gravitational force that pulls the mass less photons inward.

When the far galaxy or cluster of galaxy is very big and bright and is direct behind the center of another galaxy, cluster of galaxy or the Sun it is possible that we could see a ring all around the center situated galaxy or the Sun; that ring is precisely the energysphere of that galaxy or the Sun.

We can see multiple images of the background galaxy or star because their light pass through different layers of medium of the energysphere of the Sun or galaxy and because three points are not exactly aligned.

If we analyze a bundle of light originated from the Sun, that bundle of light does not come to the Earth in straight line but in curve line, since it was fired out while the Sun was rotating. Just like a rocket is launch vertically from the Earth. That rocket does not go in straight line to the space but in a curve, geodesic line. It is not because the attraction of the gravitational force, since the launch force overcomes to the gravitational force. That is because once the rocket is free from the terrestrial gravitational force enter to the universal rotational field which always describe a curve line until the rocket enter to the orbit of other celestial body. It would adopt the curvature of the moon, mar or other celestial body.

A bundle of light originated from the Sun to the Earth describes a curvature because the Sun launches it while it is rotating. This implies that the photons that the eyes capture come from the surface that was behind the straight point; because the Earth rotates too, it neither captures the photon right in front to the Sun but the photons that the Sun fired out 8.3 minutes before. Those photons were not come from the surface that is right straight in front to the observer on Earth.

If we analyze a bundle of light originated from a far away galaxy, for the same reason, those photons also travels in curve line. Because it travels through the Original Energy mesh which is the structural mesh of the universe; it means because the universe is not square the photons do not travel in straight line.

If that light would pass through the Sun, we would see double image of that galaxy. This lensing phenomenon has been attributed by Einstein to the effect of the gravitational force that warps space-time and attracts the mass less photons.

The true is the photons of the far galaxy light strike in the convex lens like energysphere of the Sun. We see the deflected virtual image of the galaxy first which is close to the surface of the Sun. Then we see the real galaxy which incident photons penetrate through the convex lens like energysphere of the Sun suffering reflection. That is why we see two or even multiple galaxies. The virtual image of the galaxy is always close to the convex lens like energysphere of the Sun, no matter how far away the galaxy is located. The reflected image should not be close to the real object. If we could calculate the distance of the image and the real distance of the star or galaxy, we could confirm this statement.

It has been classical the lensing phenomenon attribute to the gravitational force, that bend the mass less photons to the observer's eyes; one of the most spectacular tributes that made Einstein to become the most outstanding modern scientist. Einstein also predict the ring around the Sun or a galaxy (Einstein ring) when behind exists a very massive galaxy, cluster of galaxy or black hole. That ring in reality is the energysphere responsible to the virtual images, arcs, double or multiple images. It is also the evidence that any heavenly body possess two parts, the massive one and the energy one. It is also the evidence of the spherisity.

The Photogenesis theory has postulated that every heavenly body possesses massive part and energy part, this last one departs from the core, from the Original Energy nucleons forming the energysphere. Energysphere is the active kinetic energy part, integrating four forces: the strong force, the weak nuclear force, the gravitational force and the most important force, the electromagnetic force. Any action, reaction, transaction, transformation is the manifestation of these integrated forces. The lensing phenomenon does not prove the gravitational force pulls the mass less photons inward. The rings around any energy formation, heavenly body, black hole or supernova prove the existence of the energysphere and the sphericity of these formations. The existences of rings are independent and not dependent on the observer. The size of the ring depends on the amount of the energysphere that the heavenly body possesses.

Glare of the Sun Apollo 16 on the Moon Credit NASA

According to the author, the energy around the Sun is the medium that causes the lensing phenomenon. Different type of camera and telescope: light, X ray, ultraviolet, infrared could obtain different images of the energysphere of the Sun, other stars, black holes or galaxies. The study and knowledge of the energysphere should put the astrophysics and astronomy at the cut edge of all the sciences.

GALAXY SYSTEM

Before the beginning of the formation of the universe there was no matter, there were no heavenly bodies, there was even no atom!

At the embryonic state, there was no matter, there was no galaxy. There was no even atom!

During the hot, dense, violent vibration epoch and the big explosions there was no any matter mentioned about but the fire ball of extending and unfolding energy.

After the explosion, the new born universe expanded in accordance with the extension of the waves of all spectrums of radiations of Energy O and photogen. The universe was lighted up from the darkness!

During millions of years the scenario was predominated by photons, time that the compact photons took to unwind, stretch and extend. Time that the big fire ball took to converted to billions of fire balls disseminating to the entire universe.

Hence, where did the atom, matter, stars galaxies and clusters come from? They come from the Nucleogenesis, the formation of multiples nucleus of pure Original Energy and its transformer photogens into the nucleus of all heavenly bodies, stars and galaxies!

The Original Energy theory states that the isotropic and homogeneously distributed kinetic energy of photogens and photons inside the fire ball suffered alteration after series of explosions. The temperature suffers fluctuations by the scattering distribution of billions of small fire balls. At the extremely hot, highly concentrated photogens rotational nucleons the Nucleogenesis procedure took place. Inside the nucleons, the high frequency energetic photogens were transformed to photons, neutrinos, electrons and positrons. The nuclear reaction and the nucleosynthesis started and transformed them to subatomic particles. Quacks and gluon combined to form protons and neutrons which are spread out and get cooler. Proton, neutron and electron are bound. The enormous energy of photogen is converted to extremely heavy mass constituting the rotating galaxy.

The initial time of the universe represent the hottest time of the universe, light weight atoms were formed and converted to higher weight chemical elements, up to all kind of heavy atoms. The new formed elements were agglutinated by the rotating nucleus. The embryonic stars and solar systems were formed from the nucleus and speeded out. Galaxy was built.

The nucleus of the galaxy is under high temperature and pressure even it is lower than the initial of the universe. The embryonic immature stars, planets and moons are formed from the spinning nucleus and are spread out. Then they gather matter from the interstellar space and between the helices space where the photogens also resides abundantly and transforms to matter, until the stars got mature.

Galaxies could produce up to billions of solar systems, while they are fleeting on the sea of photogen amassing more and more matter, forming cluster and super cluster. The location and spacing of the objects are ruled and regulated by the electromagnetic mesh of the Original Energy.

Galaxy is the fundamental structure, basic unit, the principal center of energetic and material activity of the universe. The axis of the galaxy is constituted by pure Original Energy; the nucleus's disc is constituted with the transformer photogens. This energy nucleus revolves and adapts the form of a whirlwind. Through nucleogenesis and nucleosynthesis, the nucleus of the galaxy realizes the formation of all kind of heavenly bodies and matter.

At the beginning there is a small halo at the center. While the energy of photogens is consuming, the hole becomes bigger and bigger; the galaxy produces more and more stars and other objects. The spiral galaxy develops up and extends. That is the most productive epoch of the galaxy. While the energy, the fuel are consuming the spiral galaxy mature, becoming to elliptical galaxy forming less abundant stars but younger outside of the nucleus.

Depend on the amount of energy, the nucleus of the productive spiral galaxy has a "White hole", its size increase progressively while the energy of the nucleus is consuming. The non productive elliptical galaxy nucleus progressively becomes to a big hole. Once the fuel is exhaust up to its totality, the hole is as big as the size of the nucleus of the galaxy. The galaxy converts itself to a "black hole".

Our solar system was formed from the revolving Original Energy and photogens which was a fire ball inside the nucleus of the Milky Way galaxy, which was plasma. Part of the plasma suffered a differentiation becoming an independent red hot nucleus with rings of mass around.

The extremely hot nucleus became the Sun. Subsequently the mass of the rings joined together with a predominant heavier and hotter group of mass that possesses more photogens of the original energy, transforming the rings into planets. The rest of lighter mass that could not join the planets revolves around the planets, forming other rings; then they conglomerated to the hotter and heavier group of mass that possess more original energy, becoming moons. The Earth's moon was not an exception.

It means our moon was formed after the Earth had formed with the mass that revolved as a ring around the Earth.

The Original Energy Theory establishes that the galaxy is formed from the extremely hot, high pressure nucleus where the Energy O and the transformer the photogens transform to subatomic particles, atoms and matter. All kind of matter, all stars, solar systems are formed from the nucleus and spread out as embryonic bodies. These embryonic bodies keep rotating and gather more and more mass from the transformer, the photogen in the interspaces, until the heavenly bodies are matured. It means that every element of the galaxies and cluster of galaxies are formed from the intrinsic energy of the nucleus of the galaxy, according to the codes and formulas prescript in the Original Energy from the center to the peripheral in order. Galaxies are not just a random conglomeration of dusts and atoms under the gravitational effect concentrated from the peripheral to the center. Galaxy is not formed by grouping of stars and other staff under the gravitational effect, from outside to inside.

If the stars are formed by conglomeration of dusts; if galaxy is formed by grouping of stars, how are those stars, small material bodies and staff are made with such ordered system? And which is the purpose of the existence of an isolate heavy nucleus that possesses bigger force of gravity and makes all of them rotating around the nucleus? Why there is a "black hole" in the center of those conglomerated star? Why they are synchronize with the spinning nucleus?

The nucleus, the stars, all material heavenly bodies, the interspaces and energysphere around of the galaxy are inseparable united unit. Everything is formed with codes and predestinated prescription of Photongenesis, nucleogenesis, nucleosynthesis; nuclear reaction procedures and spread out by the outward energy pressure from the core to the peripheral of the galaxy.

EXPANSION

    The Original Energy theory established that the universe was originated from an embryonic energy formation. The universe departed from a spherical fire ball. There was no any matter or gravitational force existed inside; there was no any pressure or resistance existed outside the fire ball. Hence, the extremely high energetic, compact wavelength photons stretched and extended freely, accelerated with the explosions. Those photons made the newborn universe expanded with the speed faster than the speed of light. After spacing, the temperature decreased, subatomic particles, atom, mass appeared, and the gravitational force arose. The expansion slowed down. The expansion slow down drastically not just because the gravitational attractive force emerged, it is because part of the kinetic energy converted to potential energy, the decreased temperature made the extension of the wavelength increase, therefore diminished photon´s frequency. The speed of light is indirectly proportional to the amount of mass; while the universe was inhabit by atom, mass and heavenly bodies, photons traveled through different mediums changing their speed.

    During more than fourteen billions of years, the photon's wavelength of the Original Energy and all kind of photons have been extended and fragmented up to today which is the real determinant factor that makes the universe expand.

The reason that the universe is still expanding is because the kinetic energy of high frequency electromagnetic waves are still exist and are still extending. The universe would keep expanding through the electromagnetic mesh. There is no any indication that the universe would stop to expand in this extending Original Energy predominant universe.

On the other hand, our universe is not limited to the visible material boundary; there is the energysphere beyond the oldest stars and galaxies. Photogens and photons are still traveling through the mesh of the Original Energy which could be multiple size of the actual visible material universe.

The galaxies and cluster of galaxies are hold together by the photogen energy which is potential time energy. On the other hand, the recycle system retracts and compresses continually all the old stars and galaxies that are exhaust of energy. Hence, while the kinetic time energy of the Original Energy electromagnetic mesh makes the universe expand; the potential time energy of the photogen hold the material universe together. That is the most fundamental mechanism that keeps the universe expanding under a certain rate, keeps the universe ordered and stable.

The Original Energy theory proposed that it is not the gravitational force that causes the expansion of the universe. The gravitational force apparently maintains the matter, the celestial bodies in static state and in permanent order. The static state of the universe that Newton and others pioneers of the modern physic conceived should happen if the Big Bang theory is correct. Once the Big Bang explosion, inflation occurred, there was no more matter to continue forming more galaxies and more celestial bodies. The static Newtonian Theory would be still correct if it is applied to the Big Bang Theory. But is incorrect if it is applied to the Original Energy Theory where the energy is inexhaustible; it could keep transforming energy to whatever amount of mass and celestial bodies is necessary. This is the reason why the universe keeps expanding also. The universe would never stop in an absolute static state nor stop to expand. The expansion could stop until the universe reaches to its limit and the extreme high energy photons stop to extend.

The truth is the entire existence of the universe is maintained in constant revolving transformation and movement. They keep a relative relationship between them in a specific space and time, at the expense of the Original Energy, which are the forces that generate, rule, revolve, expand, inflate, recycle and maintain the dynamic functional integrity, transformation and development.

The gravitational force is not the force that expands the universe since it gets weaker and weaker in accordance square of the distance. Its intensity is inversely proportional to the square of the distance and also depends on the weight of the related celestial body: higher the weight, stronger the force; lighter the weight weaker the force of attraction. The celestial bodies move away from each other in accordance with the debility of the gravidity force which decreases progressively while the mass is consumed and exhaust.

The exterior orbit electrons are more instable than the interior orbit's electrons; the outside orbit's moons are far slower than the inside orbit's moons. The gravitational force that the Sun exerts to the distant orbit planets is weaker than the force it exerts to the closer orbit planets. The closer planets to the Sun are more compact and solid than the farther planets; they do not have or only have few moons around while the further planets have big family of moons, and even a lot of rings of asteroids.

The each time weaker gravitational force only could be responsible of the disintegration of the material unity not the ordered expansion and spacing.

Foremost, the centripetal gravitational force direction is opposite to the expansion direction. Each body's gravitational force is independent and is always attractive toward the nucleus itself. If an apple or a satellite is out of its orbit inside the solar system, they should have three destinations: smash to the Earth or the Moon; crush to any planet or its moons; or burn into the Sun. But they could escape from the solar system to the far interstellar space! It demonstrates that the gravitational forces are clumping islets, disseminated in the space with "free empty" zones between or at least a neutral zone.

The gravitational force tends to be equal to cero with the distance, between the earth and the moon, between planet and planet or between galaxies to galaxy. There are places where the gravitational force is neutral or even absent; at the exosphere there is an orbit that is the stationary satellite's heaven because they are light enough and far enough to not smash to the Earth. So the gravitational force is limited, not unlimited.

On the other hand, the unifier Original Energy is a whole force maintaining everything in equilibrium; it functions as a vehicle carrying the matter world to occupy more space making the expanding universe possible.

From the nucleus of the galaxy, the Original Energy forms every existing substance or object of the space: since the primordial plasma, up to solar system, from the center to the peripheral. The galaxy by itself is a sample of how the universe was formed. It demonstrates foremost, that it was not the grouping of solar systems that formed the galaxy, the galaxies by itself produces solar systems from the nucleus and spread them out. This is another reason of expansion.

From the nebula or supernova stars and solar systems could be formed, as in the galaxy. But they are formed after the disintegration of dead galaxies and stars that were converted into energy by black hole, which reorganizes giving place to new stars or an isolate galaxy. What seems to be a chaos, in reality is an episode of a constant transformation of the universe.

If the gravitational force causes the expansion, the universe should have a heavy capsule outside of the universe, making a solid material boundary attracts the galaxies. This is even less probable.

The equilibrium of the celestial bodies is given by the annulations of the gravity energy between them. More important is the orientation of the axis of the electromagnetic field of every heavenly body which is always different.

The existence of the solar system is maintained by the equilibrium between the Sun and the planets with their moons. The gravity force is zero between them and also between our solar system and other stars of Milky Way galaxy. Otherwise the entire solar system and the entire galaxy would be contracted and collapsed.

In reality, the equilibrium is maintained by the electromagnetic static force of the photogen, which is a constituent of the Original Energy.

During 1920's and 1930's, Friedman and Hubble determined that every celestial body goes away from us. Are we the center of the universe? We are not even the center of our galaxy but our solar system came from the center. Any productive galaxy produces billion of stars from the nucleus, and spread them from the center to the peripheral. The older stars are going away from us and we are going away from the younger ones. The younger stars are following us but because we are each time farer and farer making us feel that beyond us stars are traveling away and we are going away from the behind stars.

The same effect suffer the galaxies: the older galaxies are traveling away from us and our galaxy is traveling away from the younger galaxies and all of us are traveling away from the center of the Original Energy where we were formed.

On the other hand, the centrifugal effect of the revolving Milky Way galaxy is weaker in the peripheral, making older stars even easier and faster to "escape".

But we are not going so fast; according to the sphericity of the Original Energy theory every heavenly body rotates on ellipsoidal circle, we are not straightforward away each other! The expansion is not radial but circling.

The universe was derived from a compact, folded Original Energy which extended and stretched freely with the speed of light or even faster. The speed slowed down but the universe keeps expanding since ever. The universe would keep expanding because the Energy O is inexhaustible and keeps extending; the matter keeps recycling reincorporating energy; also because the space is still unlimited.

In conclusion the space has expanded and would keep expanding because:

Only small amount of Original Energy has unfolded, extended and transformed to space and heavenly bodies since the initial of the universe;

More amount of Original Energy could transform to less energetic photons with longer wavelength, making the universe structure bigger;

The immensity of the space permits the Original Energy to occupy more space and even to form multiple universes.

The material universe keeps transforming to energy by consuming which is very evident in stars. The other way is by recycling the matter through black holes transforming old matter to energy and later to form mass and heavenly bodies. It does not mean that the universe expands and contract periodically; it means the universe keeps transforming from energy to matter and from matter to energy and keeps stable.

Einstein recognized the Cosmological Constant was an error but actually it has been revived more and more by the Big Bang theory and the Cosmological Model, because it is necessary to avoid the disintegration cause by the endless expansion. Because the Cosmic Constant is necessary for the explanation of the repulsive force on which depends also the fate of the universe?

The Original Energy theory states that the expansion does not depend on such "repulsive force", it depends on the rate of the extension of the electromagnetic waves and the constant transformation of the kinetic energy to potential energy and vice versa. The speed of light is constant in vacuum, but is inversely proportional to the density of mass; for the same reason, mass does not travel with the speed of light as it has been established!

BLACK HOLE
THE RECYCLE SYSTEM

The Original Energy theory states that the nucleus of the galaxy is the center of formation of all heavenly bodies and material objects. The Original Energy forms the axis and nucleus of the galaxy which rotates the galaxy. Photogen is the transformer that converts the energy to subatomic particles, atom, mass, stars and all kind of heavenly bodies and objects. In this stage the nucleus of the galaxy is an extremely energetic, heavy, productive, rotating hole or "white hole" where the nucleogenesis and nucleosynthesis take place.

When the nucleus of the galaxy or a star is exhaust of nuclear fuel, exhaust of photogens, there was no more energy of photogens to transform to light chemical elements, the stars exhaust of hydrogen and helium. Then even helium and up to carbon light elements could burn out. Hence, there is no longer electric outward pressure to withstand the collapse inward force. The axis reverses the process opposite, becoming to nonproductive nucleus. The nucleus suffers series of implosions pulverizing every remnant material and converts everything to atoms. Atoms are crush into protons neutrons and electrons; positive charged protons are neutralize by negative charged electrons, the galaxy or the star become to neutron galaxy or neutron star, then converted to a black hole.

Inside the black hole any material are convert to subatomic particles, then to photons and electrons, they commence a photonsynthesis procedure. Electrons energize photons, reducing their wavelengths and increasing their frequency, higher and higher. At the same time the number of photons are reduced converting them to electrons which energize more photons becoming to extra and ultra gamma photons, generating huge energy. This process has named by the Original Energy theory as Photonsynthesis a transformation opposite to the Photongenesis process.

According to the Big Bang theory, black hole is the blackest place of the universe from where the name is acquired. It is believed that it is a singularity phenomenon where everything is compressed to infinite dense, heavy, hot point, with endless entropy to no return, even light cannot escape.

The Original Energy theory states that black holes constitute the recycle system of the universe which converts every old material to energy; it is not a singularity phenomenon converting the matter to an endless heavy super atom. It is a nucleolysis process. That is the opposite process of the nucleosynthesis of a star, galaxy or any material object. What have detected as "weight" of the black hole are in reality the implosive force converting all the matter of the galaxy, star, planets, moons, mass, molecules, atom, and every material to subatomic particles, then to photons and electrons.

Inside the black hole all four forces lose their effect and meaning, including the gravitational force. Light are photons, obviously cannot escape neither. Because all material and energysphere radiation around that originally belong to the same star or galaxy would be sucked and swirled into the hole, converting to higher frequency photons from radio rays up to gamma rays.

Electrons are separated from the atoms and rotate at the peripheral of the black hole, just as the atom. Then electrons energize the gamma rays photons; photons squeeze and compact, converting to extra gamma rays becoming to photogens. They could be energized by more electrons and be compressed even more, converting to ultra gamma rays, where only magnet persists. Then, magnet induces electric charge forming the new electromagnetic field. The old black hole is transformed. Nucleogenesis process begins again and a new "white hole" could be formed.

Any transformation is carried out by explosions. Such huge transformation should not be the exception, but they are implosions.

This is the way the black hole contributes to the production of radiation, energizing the universe. Beside the microwaves background radiation that was formed at the initial of the formation of the universe, black holes contribute continually the other part of new radiations and heat of the universe.

In conclusion: the nucleolysis and photonsynthesis processes inside the black hole could be carried out by next procedures:

1). When the star or galaxy out of fuel, it means out of photogens, the nuclear fusion reaction stop. The remnant material and radiation of star or galaxy suffers a series of implosions disintegrating and collapse. But the axis keeps rotating and gartering all kind of remnant mass, ashes, and radiations of the energysphere converting into a black hole;
2). the mass of the disintegrated star, or the binary star, or galaxy fall into the black hole and is burned up, losing their electric charge by neutralizing protons with electrons. It signified the disintegration of the matter's electromagnetic field, because magnet cannot exist along. Charge and magnet create each other and disappear at the same time;
3). atoms or any massive object are dipole, inside the matter they aligned and ordered to form a stronger unidirectional magnetic (gravitational) force. Inside the black hole the atoms are disordered by the violent vibration and heat. The magnetic potential is destroyed; consequently the electromagnetic field is destroyed, even the atom is destroyed. This is another mechanism of the disintegration of the mass inside the black hole; this proves that black hole is not a singularity procedure converting everything to an extremely massive atom where even photons could not escape. More over while the black hole whirls and mills all the mass, atoms are converted to energy, to photons, some amount of photons could escape from the black hole! But because they are high energetic photons they are invisible, they are "black" also. They could be detected only by different types of electromagnetic telescopes;
4). any celestial body has an energetic axis built up by the Original Energy, when the material is consumed, provoke by the decay of the potential time energy. The axis converted itself to black hole, instead to revolve clockwise or anticlockwise, it revolves opposite to the

original way; instead to be centrifugal it becomes to centripetal. That way, the black hole is converted to a mill that crushes all remaining material and energy. Consequently, mass becomes to composite, then to atom, then to subatomic particles up to electrons and positrons;

5). the electrons and positrons unleash a fusion thermonuclear reaction causing an even higher temperature that could reach trillions of times higher than the Sun's temperature. This temperature is lower but is similar to the temperature of the beginning of the formation of the universe where every single atom, particle, any mass is burned and transformed to radiation, to ordinary photon. After that, common photons are energized by electrons becoming to higher level energetic gamma photons. The wavelength of the photon is compacted, squeeze, and its frequency is elevated. Lower energetic photons could be converted to electrons and energize more photons. Photons become to photogens reducing common photons population.

We should know, under that extremely high temperature gluon does not glue, any kind of mass could not exist. But that is the condition required for the energetic transformation. The high temperature and the frequency of photons each time higher and higher cause the hermetic closed black hole to elevate extreme pressure that some black holes could ejects jets from the nucleon;

6). the temperature increases to trillions of degrees with violent vibration disintegrating all the mass. The kinetic time energy reaches the maximum while the potential time energy becomes to cero. The temperature is extremely high at the beginning of the phase transition, but later entropy enters in action of the transformation, temperature also diminishes;

7). the axis of the star or galaxy is a magnet bar; it keeps rotating while it grinds up any existence. The entire structure of the old star or galaxy collapses without the electric repulsion support. It means it is not the gravitational force that makes rotational black hole; gravitational force concentrate masses with radial way not revolve or crush; it should construct a super atom by compressing, but it is not because the gravitational force disappears with the mass. Therefore it is not a singularity process;

8). what it implies is, the black hole gather and conglomerate all material and energy. The radiation, the energysphere that form the star or galaxy could not escape which equivalent that the light could not escape! We should know the star or the galaxy consume their body converting to energy since their formation. When the fuel is consumed, the entire star or the galaxy had almost converted to radiation which is integrated to the energysphere.

After that, the Nucleogenesis processes begin converts the black hole into a "white hole", transforming into a new energetic nucleus of photons O and photons E. The nucleosynthesis begins again. New stars or galaxy would be born if the black hole is big enough or simply all the energy integrates to the universe. Inside the black hole, would not have another explosion until the formation of supernova. Extra energetic gamma ray burst explosions could happen;

9). all the electrons of the atom are apart to the periphery of the black hole, they do not shine; photons are twisted and compressed to higher frequency, they do not shine neither; the explosion is implosion, it does not shine. That is why the black hole is so black! At the change of phase explosion occurs releasing different range of radiations and matter, becoming visible supernova, then it shines.

If the black hole is a singularity manifestation, reducing galaxy to a super heavy neutron atom forever, it would has an immense impact of gravitational force that alter the homogeneity, space-time around and its disequilibrium observable everywhere. The universe instead to be homogeneous it would be clumpy everywhere. Fortunately, when a star or galaxy is out of fuel, most of the matter had already transformed to kinetic energy dispersed and keep into the energysphere delimited by vacuum. The rest becomes to ashes, later, ashes also transform to energy and then integrate to the black hole or the universal energy. Some of them would become to supernova and new heavenly bodies again. That is the way the Original Energy maintains the ordered equilibrium, homogeneity and conservation of energy of the universe.

The black hole by itself proves the gravitational force is not the force that could cause the singularity up to no return, neither the force that integrates the universe. The mass dependent gravitational force simply disappears when the mass disappears in the hole! What remains is energy.

In this sense we should distinguish the productive galaxies from the nonproductive or recycling galaxies. The nucleus of the productive galaxy is extroversion energy; realizing nucleogenesis and nucleosynthesis processes; while the nucleus of a nonproductive galaxy (black hole) is introversion energy, realizing nucleolysis, photonsynthesis processes. The nucleus of productive galaxy should not be called "black hole", it is constituted by pure Original Energy axis and photogen. Meanwhile we could use that name for the nucleus of nonproductive galaxies or stars in their final step which would be recycled.

The black holes difficultly could be the routes of communication of our universe with the multiverse. It should not conduct the energy to other universe breaking the integrity of our universe, breaking the energy conservation law. The black holes would converted entirely to radiation energy and be integrated to the Original Energy through photogens energizing the universe. Nucleogenesis could begin inside the black hole!

Anyway, matter only constitutes four per cent of the Universe. We wonder: Which would be the mechanism of the singularity to make the rest ninety six per cent of the dark matter and dark energy to a super heavy atom, to an extreme dense point? Moreover, during the beginning, during the Big Bang, which was the mechanism to disintegrate such immensely heavy compact mass, with whatever big explosion, to the degree of pulverization in seconds? What kind of force could break down such immense gravitational force? Gravitational force by itself should not be guilty for such event!

The Original Energy theory affirms that the black hole is the recycle system of the universe, the final stage of star or galaxy or any material where the remnant matter and energy of the heavenly body are converted to electromagnetic waves which are compressed, squeeze and energized by electrons becoming to extra gamma rays up to photogen radiations or even to ultra gamma rays which is the Original Energy. The action of a black hole is a Photonsynthesis process.

In a change of phase it could suffer an explosion, it might become nova or supernova. The energy would be integrated to Original Energy and be used to transform to new star, galaxy, heavenly bodies. The energy also might permanent for the maintenance of the space.

The Original Energy theory has insisted that any heavenly body consist in one massive part and other energy part. Energy is emanates from the core and constitutes the energysphere which forms the spacetime.

While the energy and matter are consuming inside the nucleus of a star or galaxy, the kinetic energy increases in the energysphere which could form a ring of energy around but is kept inside the energysphere around the heavenly body. When the heavenly body is out of fuel, the mill of the black hole is formed, the rotating axis suction every remnant material, but the most important phenomenon that occur is the electromagnetic axis of the Original Energy attract all the photons and electrons of the energysphere to form a new nucleus! That is why all the matter collapse and fall inside the black hole; but the primordial part is the energysphere which consists with photons and electrons; they are attracted to the magnet axis of the black hole also. This is the reason that electrons and photons, event light could not escape!

Nucleogenesis and Nucleosynthesis processes are repeated, making possible the transformation of the kinetic energy to potential energy. Through Nucleogenesis, the nucleuses of new stars begin to form inside the black hole. In the change of phase explosion occurs. Depending on the amount of energy and the amount of mass that the black hole possesses, the black hole is converted to a luminous nova or supernova. Energy, all kind of elements, and even new stars are spread into the supernova energysphere space, which could reach up to the energysphere of other heavenly bodies.

Ring of the Black Hole.
Credit: X Ray NASA/CXC/MIT/Rappaport et al
Optical: NASA/STSc

ENERGY FIRST

Since the beginning of the formation of the universe, it was the energy that transformed to matter, not the matter that converted to energy. It means the universe was formed from the Original Energy which was constituted by ultra energetic photons. The universe did not derive from a primeval super atom or from a super concentrated mass to the degree of singularity. Even less probable originated from nothingness; nothingness only could form nothing!

The new born universe was constituted exclusively by Original Energy and photogens, the extremely compact wavelength embryonic photons, precursors of ordinary photons. Those photons suffered extreme high temperature and pressure, agitation and vibration, collision and friction. Hence, the entire universe was a sphere of heat radiations. During photon energy dominated epoch, photons travel faster than the actual speed of light inflating, expanding and forming the new born universe.

Today, the statement of "nothing travels faster than the speed of light" is still valid, because they are two different entities and different scenario. Einstein referred none of physical objects travels faster than the speed of light. The Original Energy is not a tangible object, its action could be transmitted simultaneously everywhere as a whole when an action takes place, even at the other side of the universe. That's why humankind has only been able to unravel few of the physic laws of the universe, because it limited to physical objects, not high energy level. On the other hand, the precursor photons traveled in an absolute empty vacuum, not as the actual photons that got to travel through different mediums, suffering refraction, reflection, diffraction and interference. They had stretched long distance and long time ago. The speed of light has changed and kept constant in vacuum now.

The analogy between the origins of the universe with the World War II atomic bomb explosion is unacceptable, since the bomb was matter that was converted into energy annihilating the matter; while the creation of the universe was opposite: the energy given rise to the matter: matter is always created from energy.

The matter accumulates and conserves energy, making the energy tangible, visible, touchable. The energy concedes life and spirit to the matter. The matter exists because it contains energy. When it loses the energy, it disintegrates, vanishing, transforms and even explodes completing the cyclic universal law of birth, living, develop, death, recycle back to energy state and to be reborn.

We are living being because the Earth is a living being; the Earth is a living being because the Sun is a living being; the Sun is a living being because the Milky Way galaxy is a living being; the Milky Way galaxy is a living being because the universe is a living being. We all are because we possess energy, the Original Energy in a macrocosmic concept.

Matter without energy is death. The most exact definition of death of any object or living being is exhaust and emptiness of energy, of Original Energy. It could happen to people, to any living being, to galaxy, to star, to planet, to any existence.

But death is part of life, it closes the cycle of the transformation; from living being, atom, stars, galaxy up to the universe; transformation is the scenario of a battle from photogenesis to entropy, from ordering to chaos, from the "white hole" of a productive galaxy to the black hole of a galaxy in agony, from integration to disintegration.

The energy could transform to matter; the matter could convert to energy. The live functional state of the matter is with energy. That's why energy should be considered as the fifth state of the matter, besides the liquid, solid, gaseous and plasmatic states.

In reality, matter or mass is a manifestation of electromagnetic field. Mass is constituted by atom; atom is constituted by proton, neutron and electron; proton and neutron are constituted by subatomic particles, subatomic particles are constituted by photons and electrons. They all spin and formed an electromagnetic field. Atom has its electromagnetic field, planet, star have their electromagnetic field, because they are built by photogens and photons. So energy is a metamorphosis of matter.

Mass is potential space-time energy, through time it decays, converting to kinetic time energy. The relative space, time that the mass occupies would reduce progressively. The entire potential time energy could become to kinetic time energy. The space it occupied would be exchanged and filled up with energy. In this case, kinetic energy reaches its maximum which is photons with the speed of light. So photon and mass are interchangeable, just like kinetic time energy and potential time energy. Photons could be kinetic time energy and potential time energy because photon itself is potential and kinetic energy at the same time.

Energy always exist, mass could disappear but energy never. That is the "secret" fundamental of uncertainty; this is the double personality of photogen.

The pretension to reach the discovery of the origin of the universe by searching the endless subdivision of nuclear matter, not only technically is impossible but it would not get to a formation of a mini universe. More probable energy would result as the final state. The ascendant study from mass to energy has discovered successfully a zoo of subatomic particles; the descendent study from ultra and extra energetic rays to matter, (higher than gamma ray), might achieve even greater result.

The Original Energy Theory establishes that every existence of the universe derives from Original Energy, therefore from photon´s Energy. Neuronal cells are not the exception. Central nervous system, peripheral nervous system and sensory receptors form the most sophisticate structural and functional complex of any organism. In the case of human, the interaction of receptors (information), central nerves system (connection, analysis and decision) and the peripheral nerves system PNS (execution) control the whole body.

The central nervous system is made with the brain and the spine cord. The brain is formed by the forebrain, midbrain and hindbrain. The forebrain consists of the cerebrum, thalamus and hypothalamus. The midbrain is the tectum and tegmentum. The hindbrain or brainstem is made by cerebrum, pons and medulla.

There are three types of neurons and their basic functions are:

1). Sensory neurons with long dendrites and short axon which carry message from sensory receptors to the central nervous system. Their functions are receiving sensory input from internal cells or external environments;

2). Interneuron that connect neurons to neurons which are localize only in the central nervous system. Their function is integrating the input, integrate the analysis between neurons and send out the orders;

3). Motor neurons which respond to stimuli. They have short dendrites and long axon that transmit messages from the central nervous system to the peripheral muscles, organs or glands.

Receptors are sensors that detect changes of internal or external stimulus. Sensory receptors as auditory, visual, tactile, olfactory and taste gather information from the environment outside the body. The inputs could be all kind of information that the sensory receptor capture, including the internal autonomous system receptors of sympathetic and parasympathetic system functionality from inside the body. These afferent information pathways are converted to signal and send through the peripheral nerves to the spine cords which speed the message to the hindbrain. The hindbrain selects and distributes the message to the midbrain and this to the forebrain.

But which is the meaning of all these sophisticate structure? How could we explain such diverse cerebral activity as intelligence, reasoning, intuition, language, memory, conscious, unconscious, thought, dreams, mind or soul? If we have a palace it doesn't mean we have a functional empire. Cerebral mass is even a lot more complicate but by itself cannot work like the most perfect computer!

Neurons like all other cells are charged object. Hence, when any part of the body receives a stimulus, the sensitive receptor's electrons are excited, and then the sensitive cells vibrate and generate electric currents. Neurons have positive charge outside the plasma membrane and less positive charge inside the membrane, making a negative charge gradient inside, which is a resting potential. Sodium ions are more concentrated outside the membrane while potassium is more inside. The action potential result when stimulus produce change of the polarity of the membrane causing the concentration of sodium-potassium reversed. Then inside the membrane is positive and outside is negative. The sodium-potassium pump restores the resting potential membrane charge by pumping sodium out and potassium in. This action is carried out along the neuron body; it is an electrochemical action where the dendrites and axon membrane suffer depolarization and polarization. It means the message is transmitted by electromagnetic waves.

The neurons cells carry out all kind of cerebral activities. The dendrites capture the ascendant impulses; depend on the kind of message the neurons respond with an order; the axon conduct out the messages through the synapses where the chemical neurotransmitters send out the descendent impulses. Impulse message travels within the neuron as electrical potential; electrical potential induce magnetic potential. Hence, messages are electromagnetic waves.

The connection from neuron to neuron is through the synaptic cleft where the neurotransmitters are localized inside vesicles. Message could be transmitted or inhibited by transmitters. Transmitters could be composite acetylcholine, norepinephrine, adrenaline, noradrenaline, dopamine and others hormones. Hence, the synapses only contribute in the selection, excitation, inhibition, connection, communication and distribution of the impulses between the neurons network. The neurotransmitters do not travel along the neurons dendrites or axon; they are limited inside the synapses cleft. The real work is made by the body of neurons with impulses.

We know the brain, the nervous system structure, the way the impulses formed and transmitted but it is still not clear how and what made our mind! Mind should be formed inside the neuron and integrated in a group of neurons.

Any time, in any circumstances when electric current is created magnetic is created simultaneously, therefore the electromagnetic field. The impulses are formed by the difference electrochemical gradient of sodium and potassium ions on both sides of the cells membrane but the messages that the movement of impulses carry along the dendrites, neuron body or axon or any kind of cells are electromagnetic waves. Therefore memories, thought, languages, identity, consciousness, reasoning etc. are parquet of electromagnetic waves!

Therefore the neuroreceptors capture the information which are photons resulting from releasing energy of the exited electrons of the objects. The photons are transmitted via afferent to the spine cord, then to different neurological levels up to the cortex. These electromagnetic waves are converted into biophotons which are awareness. The awareness biophotons are transformed to consciousness and are sent to different neurological centers. The neurons respond analyzing and making decision, others biophotons are carried out as an order. Peripheral neuron's electrons by vibrating release other signs that are messenger biophotons. Messengers carry the order and reach to muscles, visceras or organs via efferent; the electromagnetic waves excite the neuroterminations of the muscles, glands, visceras or organs. They respond executing the order by releasing energy which is ATPs the combination of photons and electrons.

Structure and function are integrated by the energy which is electric and magnetic potential waves. Both have been able to be detected by electroencephalography or magnetoencephalography. The known electrocardiography, electromiography are another manifestation of the electromagnetic field of the body.

Everybody knows the miracle potential that the seeds are capable to develop. We all could know its weight, its amount of matter, the four forces it has are insignificant as well. Why could they make such a transformation? Why a microscopic single ovum and a tiny spermatozoid could make such complex structure and functional mankind? Simply, that is because the seed, the spermatozoid possess Original Energy. Not because their matter suffer a kind of Big Bang explosion inside the earth or inside the mother's uterus! Neither just because they possess DNA and RNA, they all need the energy to start to function and develop! Without energy DNA stop to work and death overcomes.

In the case of the spermatozoid and the ovum, they need the energy attraction first, then the guidance of the messenger RNA, physical contact, chemical reaction and join both parts of twenty three genes, before the DNA enter in action.

With our receptors, we collect all information necessary and analysis them before we make decision; we made the ideas before we enter in action; we design everything with our mental energy before they are materialized. The cerebrum is an energy machine, energy storages in the neurons and the hippocampus. It is possible, when a stimulus takes place, the hippocampus release an extremely small amount but extremely ordered synchronized energy that distribute to the entire cerebrum. Then the correspondence area of the cortex selectively responds and releases codes constituted by biophotons. After that, the molecular chemical process takes place. The smartest way to storage information is with codes and messages, with energy waves; not by storing voluminous molecules or matter.

If the entire molecular theory is true, if memories, mind, thought, language, consciousness, soul are protein's molecules, we should have a craniums and cerebrum hundreds or thousands times bigger than the one we have. Then, we should have a big body to support such a heavy head, as a dinosaurian.

On the other hand, if all cerebral processes take place with molecular chemical reaction, it would generate very high temperature that could burn up or cook up the entire central nervous system. We should have a complex radiator to cool up such elevation of temperature.

Memory is considered as the connection of neurons circuit at the synapses with the action of serotonin and the selection of different protein composites. Definitely it is a lower level of the memory processes. Memories as other cerebral functions are neuronal cells intrinsic power, all of them depend on their structure, the amount of neurons they are connected and the energy they possess.

The synapses play an important function in connection between different groups of neurons. Those neurons integrate input information, retrieve codes data, integrate ideas and send out action order as codes through axons to dendrites. The synapses cleft is formed and limited by the presynaptic neuron dendrites; the glial cells, the astrocitos dendrites and the postsynaptic neuron dendrites. The presynaptic neuron sends the electric impulse to the synapses which excites the cleft; the astrocitos segregate transmitters inside the synapses capsule; the transmitters excite the postsynaptic neuron; then the transmitters are destroyed and are recycled immediately. Only electromagnetic waves travel through the neurons not any chemical components!

All these codes are biophotons, extremely organized, synchronized with each neurological center.

The electrical afferent and deferent impulses of the peripheral nerves, the electroencephalographic electric manifestation, chemical transmitters of the synapses and the molecular formation of the central nervous system activity are only third and second level reaction data, the highest level should derive from neuron's energy.

And how the encephalic neurons communicate each other to integrate ideas? How the memories are stored and retrieve selecting out each time we need any information? What constitute reasoning, auto perception and consciousness? They are codes and messages derived from the deep located original energy. They should not be subatomic particles or atoms, even less possible voluminous molecules or composites of molecules! They are compacted electromagnetic waves of the Original Energy.

The reality is all neurological activity functions with electromagnetic waves at the highest level of neurons and even at the peripheral nerve system. Most of the chemical molecules as neurotransmitter and other substances are limited in the synapses capsule and conjunctive glial tissue.

One of the most concrete examples that demonstrates, our brain as our Earth, our Sun, and our universe function with electromagnetic energy is our self consciousness. Our eyes detect directly all color electromagnetic rays which enter to the retina and reach up to the occipital cortex and then are distributed instantaneously to others centers to integrate the surrounding data; with these detectors we can perceive and be aware of all surrounding conditions. We have an electromagnetic field all around our skin; we can feel even without seeing what is behind us. Combining others receptors that send resonances information to the cerebral neurons, the brain analyzes, retrieves experiences and form the consciousness which are electromagnetic waves. Then selectively the involucrate area responds and sends the action order which is group of biophotons to the executive organ or muscles.

The most familiar that all activity of life is the activity of photons is the heart. The heart is built by muscle but it functions as an electric generator, the electric induces the magnetic and we capture its electromagnetic activity with the electrocardiogram! It should be named as electromagnet cardiogram.

We can affirm: every biological being on Earth derived from the action of light and all spectrum of radiation of the Sun combining with the Earth's electromagnetic fields which are photons. Our brain is a small universe, function with interconnectedness resonance electromagnetic field, constituted by weak frequency biophotons. The magnetic field generated by the neuronal electric activity is as small as millionths times the Earth magnetic field.

Our brain is made up with cells, neurons and nerve fibers that function with low frequency photons which are biophotons. All our perception is converted in waves of biophotons. All our cerebral activity is constituted by the electromagnetic field, which forms a complex network with diverse center of neurons, where the interconnected biophotons travel as fast as the speed of light, constituting all the mental activity. Directed by the Energy O they make: mind, consciousness, language, intuition, reasoning, creativity, learning, intelligence, memory, soul, even dreams.

LIGHT IS PHOTON
PHOTON IS LIFE

The solar radiation is the motor and fuel of every activity on Earth. Sun's energy directs the weather system; the variation is caused by the impact of the heat and the tilt of the axis of the Earth. Combining with the geomagnetic field, extrastelar radiations and the geographic conditions as polar, arid, subtropical, tropical, mountainous, forest, desert etc. making different climates and stations.

All kinds of radiation could reach to the exosphere, but most of the Sun energy reaches the Earth surface as visible light and infrared radiation. Eight percent is ultraviolet which is prevented by the ozone of the stratosphere. Others harmful radiations are prevented by higher levels atmosphere as ionosphere and thermosphere. Anyway the solar radiation could reach the Earth surface after they become to lower frequency radiation while they pass through different layers of the atmosphere. The solar eruptions periodically cause significant variation on the weather.

Energy changes take place during the penetration of the Sun light in the exosphere, ionosphere thermosphere and stratosphere before the **climate** activities physically enter in action in the lowest layer which is the troposphere. It means hurricanes, tornados, typhoon formation; any climate activities, the transformation of the energy take place on those higher levels of the atmosphere, before they are formed physically in the troposphere.

The invisible spinning energy axis is formed before the storm, tornado, hurricane materialized. The hurricane is not just a mechanic action suctioning the salted sea water to the sky. It is a physical chemical procedure converting the salt water into sweet water before rain. It requires enormous quantity of energy to transport the water from the sea to the continent made by electrostatic attraction. Hence, weather is just a physical visible phenomenon; the real force is the action of the Sun radiation energy, the action of the Earth energy, the action of photon.

The incessant change and transformation of the weather and climate make up a suitable environment for the development of all kind of life. The diversity of living being depends precisely on the constant changing of the climate which primordially is the action of the Sun energy.

Solar radiation is made up by electromagnetic waves with different ranges of photons. The atmosphere and the surrounding geoenergysphere protects the Earth from harmful radiation such as extra energetic gamma rays, gamma rays and X rays, and the solar wind subatomic particles that only reach the highest atmosphere level. But these radiations could ionize and collide with atmosphere particles, transferring their energy becoming lower frequency photons, as visible light rays and lower ranges rays and then they could enter the troposphere reaching the Earth surface.

The Earth is a live planet because the action of the energy of the Sun combines with the energy from the core of the Earth producing biophotons which are the real constituent of every life.

The visible light only can penetrate superficially the ocean and the continent. But the longer wavelength radiation could penetrate deeper ocean which might take place in the photoluminescence function of grand variety of creatures too. The photoluminescence is made by chemical component, luciferin-luciferase reaction converted to photons, most of them under the deep sea. Somehow the invisible long wavelength as radio, microwave and infrared radiation are converted to visible luminescent photons through Photongenesis transference of energy from photon to electron and from electron to photon up to very low frequency biophotons.

In all the photoluminescence ATP energy are involved. ATP is made by photons and electrons; hence photoluminescence is the partnership between photons and electrons! The ocean is the biggest reservoir of

Sun heat which contributes in all life style of the ocean and the climate of the Earth.

The electromagnetic radiations take action in the photosynthesis processes almost at all kind of plants by transforming photon's energy to chemical energy, into sugar and oxygen; they take part in photochemical reactions of animals also.

Photosynthesis is a process where the sunlight energy is converted to usable chemical energy. It involves several protein complexes of the chloroplast.

The photon magnetic and electric energy bi-structure excites two electrons which energy potential gets higher. The electrons are transferred to a protein mobile carrier which picks two hydrogens up. The electrons are replaced by two split molecule water; one molecule of oxygen is produced as secondary product. Two more electrons are transferred to two more protein complexes, the third one uses photons to energize the electrons, then the energy is used to produce ATP. The actions of photons and electrons are fundamental.

The ATP energy is used to make sugar and others products to feed the plants. ATP is the fuel used by all living being.

Under the action of the energy of photons, the final result of photosynthesis is: six molecules of water plus six molecules of carbon dioxide produced one molecule of sugar plus six molecule of oxygen. The inorganic carbon dioxide is converted into organic sugar and oxygen by photoautotroph organisms. The energy comes from photon and continually supplies to electrons.

Photoluminescence is used broadly on camouflage, illumination, communication, repulsion, sex attraction, defense by sea living being just as us using biophotons in all life activities.

For more than four billion of years, living being have been magnetized and electrified. Cells, muscles, bones, especially central nerve system work as electromagnetic system. During day time we work with photons of the Sun; during night time we work with photons of the electromagnetic field of the Earth. The nose is not only serving for breath but also is used for detection of variation of temperature, humidity, compass orientation of North Pole for navigation, especially for birds. Most of the animals gather, sniff, distinguish and analyze food with the nose and tangle.

When we go out or enter a cold place our mucous membrane of the nose reacts sneezing immediately with the change of temperature. The pituitary gland is localized behind the optic chiasm of optic nerve, behind the olphatory complex. The pineal gland segregates magnet crystals which are essential for orientation and navigation.

Trees and forest vegetation compete Sun light growing up or are leafy; also compete absorbing the magnets from the Earth by growing abundant and deep roots. Fresh vegetable, spring water, mineral and the nutrients that come from the Earth are magnetized. They are healthier while we eat or drink them fresh, otherwise they are neutralized.

If we analyze function by function, any behavior of the organism of all living being, we reach the conclusion that all are dependent on the Sun light or Earth electromagnetic field. It means all the living activity is the conjugation of photons and electrons.

According to the Original Energy theory the solar system derived from the energy of photons; the Earth derived from photons; the entire universe derived from photons; we should not surprise life derived from the energy of photons. The question is why in other part of the solar system there is no life? Why in other part of the Milky Way galaxy we have not discovered life yet? Until today we have not realized that there is life in other part of the universe. Foremost we have not encounter intelligent life in other part but Earth! Then, the research of the specific condition that made possible the development of life and intelligent being on Earth is fundamental. Life could be found in other planet of other solar system with similar conditions as the Earth. We won't find similar life in other part of our solar system because the Sun light and electromagnetic radiation that other planets or moons in our solar system they receive are not propitious for the development of life. Or only could form stranger creators for science fiction films.

The formation of life occurs on successive long period of time of evolution, transformation, adaptation, even mutation. Life appeared under extremely complex conditions of the combination of radiation, light and especial condition on Earth. Photosynthesis is just a very basic procedure to provide energy for life. The alimentary chain by itself demonstrates the complexity, dependence of every species in the formation and evolution of life. But all of them, any form of life are linked with the suitable terrestrial electromagnetic field which made possible for photons that come from the Sun and outside of our solar system transform to suitable biophotons. That is why light is photon, photon is life on Earth.

ORIGIN OF LIFE

The Original Energy theory states that the universe was formed from the ultra high frequency energetic, compact wavelength photons and photogens which carry the codes of every existence on Earth, Sun, galaxies and the entire universe. Photogens are the most fundamental elements of the universe from which all the common photons, the subatomic particles as quacks, gluons, and the entire zoo of subatomic particles were formed. Therefore also proton, neutron, electron, mass, heavenly bodies and the entire material universe derived from extra energetic photons.

By the same reason, life as plants, animals and intelligent being were originated from the Photongenesis procedure of photons. Through photogenesis the mutual generation between photons and electrons processes, photogens were converted to longer wavelength photons progressively. It is probable that a division of photogen messenger with the codes of life was converted to very low frequency photons on Earth.

The solar radiation which is the spring of everything is photons, they acts on organic molecules converting them onto primeval unicellular plants which had the faculty to realize photosynthesis. After that, the unicellular bacteria were formed. Unicellular plant and unicellular animal formed the symbioses environment. This implies the mutual supply of oxygen and carbon dioxide and even more important the administration of energy ATP. That was and is the way that the coded biophotons, highly ordered and synchronized, low frequency photons appeared and that is the origin of life!

Obviously organic composites were necessary before the cellular formation, just as the inorganic composites were necessary. In a further step, the diversity of the evolution of different species through photosynthesis, alimentary chains; recycle of gas and organic material constitute the complex communality of living being.

On Earth and on everywhere of the universe the only one fundamental element that exist since the beginning of the universe is photon follow by electron. Not even hydrogen or helium but photon, less probable a composite of extremely heavy atom. Any existence in any point of the universe contains photons. It means everything derives from the same trail which is photon! Moreover, the intelligent being derived from photon because photon is an intelligent being!

Photogens and biophotons possess codes of every existence which consists in different ranges and arrangement of electromagnetic wavelength and frequency.

Even the temperature of the birthplace of everything is the same, one part in 100,000 everywhere in the universe, because temperature is radiation, kinetic energy built by photons. Temperature has been a determinant factor for any formation in the universe, even more important for life formation on Earth.

In a further stage the subatomic particle were formed constituting proton, neutron and electron. Ones the atom is bonded, under high temperature and pressure it is not difficult the combination for light chemical element to form heavier ones and later the composites.

The optimal magnetic, electrical, chemical, physical, geographical, atmospherical, climate, temperature conditions of the Earth; the precise location of the Earth on the comfort zone of the solar system and the comfort zone of the galaxy, made possible the formation of oxygen, water, amino acids, nucleic acids, glucose, carbon dioxide, nitrogen, then ARN, ADN and other vital components, first for plants then for animals. But without the action of photons and the suitable Earth electromagnetic field, the inert inorganic material could not be able to transform to organic material, suitable for life. Neither possible to recycle the toxic gases and convert them to breathable gases cyclically. The symbiosis between plants and animals exchanging carbon dioxide for oxygen make them dependent each other.

The ability as messenger of photon makes possible the apparition of the vital messenger ARN, which makes possible the transference of every life codes. The formation of ADN makes possible the transference of genes from generation to generation. Ones again, any transformation from simple light chemical element to heavy element, from elements to composite, from composite to molecules, from inorganic to organic needs the energy of photon, needs the messages of photons, needs the partnership combination of the action of photon and electron. All of them are the result of photogenesis. These are the most fundamental mechanism of the formation, transformation and evolution of life.

Earth as others planet has its individuality; its topographic situation is unsurpassable; its temperature is isolated by the vacuum; its composition is unique. Seventy per cent of the Earth is water; seventy per cent of the human body is water. It is not just fortuitous. It is obvious that the favorable condition of Photongenesis made possible for biogenesis.

Yet, water which is hydrogen and oxygen; air which is nitrogen and oxygen constitute the most fundamental elements of life. Hence, life was originated where these elements are continually reacting, combining, transforming and recycling under the bombard of the Sun and Earth radiations. It implies life results from multifactor processes as photogenesis, photosynthesis and metabolism where the actions of photons and electrons conditioned an adequate temperature and environment. Hence, life emerged close to the water, close to the air, at the riversides and the sea coast where vegetation and fauna developed and could realized all kind of biological processes as metabolism and photosynthesis.

Life was originated on Earth and not come from other planets or solar system as meteorites; the meteorites and cosmic radiation, supernova might contribute supplying some material necessary for the formation of life but it means never we are descendents of meteorites. Otherwise we should have company today in other locality and should be able for human being to inhabit those places of the universe, where we supposed come from. The Moon and other planets are full of craters caused by meteorites and there is no a single cell on them has been developed. Supernovas have been proved to contribute some heavy elements which is not mean neither life come from such violent explosion.

Meteorites that reach the Earth suffer collisions with the cosmic radiations, with different layers of the atmosphere of the Earth and crash with the surface of the Earth elevating extreme temperature. Most of them disintegrate and evaporate by friction and collison. Consequently, life if any that comes with meteorites would not survive!

Life comes from light, naturally from the Sun; life was originated on Earth principally by the transformation of photon and electron to week energetic biophoton. Then by the combination of biophotons and the specific, intrinsic conditions of the Earth! One of the most outstanding factors is the electromagnetic field of the Earth which not only protects the Earth against the harmful radiation but also supplies the suitable photons to become biophotons.

As the common photons, biophotons have the faculty of messenger transmitting orders or energy for biochemical, biophysical and biological reactions where biophotons are absorbed or released with different frequency. Therefore any kind of life absorbs and emits biophotons. The emission of low frequency biophotons by all the cells of an organism form the electromagnetic mesh making them organized, synchronized, communicated, functional live unity.

As it has been emphasized, photogens are localized on the nucleus of the atom and cells, especially in the reproductive cells. Therefore, biophotons that derived from photogens are found on those localities. They intercede on reproductive procedure, cellular mitosis; transformation and combination of genes, hormonal regulation, metabolism, blood circulation, gases interchange circulation, cerebral activity and all kind of biological activities.

Birds, fishes and grand variety of animals possess organs that use electromagnetic waves for navigation, communication and orientation, defense and offense. The impeccable synchronization and organization on their formation on flying and swimming are the function of biophotons. Moreover fishes and other organisms of the deep sea posses spectacular photoluminescence, they use biophotons for illumination, attraction, defense, sex, communication and other activities. The newborn seals cannot see but they distinguish their real mother from the community by smelling and hearing their voice where the particular particles and waves are converted to biophotons.

The neuroreceptors of the skin, mouth, tongues, and nose detect different molecules or grade of temperature and substances by using different frequencies of waves, especially on reptiles, dogs and other animals, even human babies. Photoreceptors of the retina, receive directly all spectrum of colors which are electromagnetic waves. They are transmitted by the optic nerve to the genicular body, optical radiation fibers up to the occipital region of the cerebrum. The audio receptors of the tympanic membrane and middle ear receive the acoustic waves. All these waves are converted to biophotons by the receptors neurons.

The millennium therapeutic method of acupuncture states that we have an energy circulation system which has not been confirmed by any scientific method. If the existence of biophoton in our body is out of doubt and its energy circulation network form part of live system, we are mostly talking about the same subject! The meridians are the pathways of the energy circulation! Taichi, yoga, meditations, even the conventional method of exercise have demonstrated the activation and concentration of biophoton which improve or even cure diseases of the body! Hence, the existence of energy circulation system should be out of doubt.

If some of the predictions of the visionaries and psychics were truth, how did they make them? If we make plan predicting the results, how we make them? Definitely they were made with photons, Original Energy. The mind with the original energy could travel back to the past or toward to the future. This is the action of the biophoton.

Climate is a metamorphosis of the energy. From born to die of the star or galaxy, from plants, bacteria dolphin, up to human kind, all of us have been part of the process of transformation of the energy. The huge transformations on Earth as glacial era, deluges are consequence of huge changes on the Sun or Earth electromagnetic field.

The same atom of oxygen, hydrogen, carbon, nitrogen or others elements have been part of all materials million of billions of times by reusing them. The objective matter is a physical manifestation of those transformations. Without energy it would not have any transformation of the matter! Life is one of those transformations from dust to dust, from abiogenic to biogenic.

Biophotons are the fundamental element in the energy circulation of any live organism. Life, spirit, all our cerebral activity are formed with the energy of biophotons.

Human kind has believed to be the superior being over any other kind of living being because human kind has the faculty of language (sound waves) to communicate each other. If we knew the inferior being use broadly biophotons, long wavelength electromagnetic wave for communication from ocean to ocean, from continent to continent which human being are just begin to learn how!

Earth is a huge magnetic sphere that changes since its formation. It magnetizes and electrifies every matter and every living being. Every cell, every organ, of every living being from generation to generation had been dependent on the electromagnetic transformation. These transformations determine different steps, different spices of the evolution. It means we are direct product of the Earth, the Sun, the galaxy, even the universe electromagnetic field evolution!

Each individual has a specific biofield with proper arrange of frequency and wavelength, but as all matter or living being is made up of electromagnetic field with different ranges of photons, they could be influenced and interfere by any other electromagnetic field and even cause damage becoming illness.

Electrical currents flow in living being, creating magnetic field which could extend outside the organism and could be influence by external magnetic field also. Electrical currents of the brain creates magnetic field that extends outside the body which could be measure and analyze with the electroencephalography. Any living being is surrounded by magnetic field. The emission of biophotons by living being has been demonstrated, we can detect that weak or ultra weak frequency photons glimmering out of the human body, just as any heavenly body that has energysphere, the human body has energysphere surround. We are not saint but we possess aura.

Any cell, especially neuron is an electric charge unit which forms an electric field, its action induce the formation of the magnetic field, together constitute the electromagnetic unit. Any cellular activity finally is in fact an electromagnetic wave of biophoton.

The Photongenesis theory postulates, life is a process of absorption and emission of photons. The entire life consists in the cellular activity where the transformations of photons are the most fundamental process. We capture photons from light, sound, nourishment, air, water and surrounding; photons are the elements that maintain us alive, through metabolism, oxygenation, blood circulation and biochemical conjugation with electrons procedure. They are transformed to lower frequency photons which are ultra week biophotons. Then we utilize the energy of photons for any and all our activities, emitting ultra week biophotons glimmering while we alive which constitute the real spirit! The entire mental activity consists in the activity of photons. Moreover the equilibrium between centropy and entropy (organized and disordered) of the energy of our body determine health or sick, live or death.

Any electric or magnetic cellular hyperactivity or hypoactivity would manifest as illness or malaise. Cancer is a cellular electromagnetic hyperactivity; diabetes mellitus is a cellular electromagnetic hypoactivity. Long before and on the book "Ecos de Reflexiones", the author has pointed out that beside the central nerves system, blood circulatory system, breathe gaseous circulatory system, endocrine hormones circulatory system, the body has an energy circulatory system that control the electromagnetic energy circulation of the whole body. That is why the regulation of cellular electromagnetic activity up to the equilibrium could cure our body. One of the oldest therapeutic methods, the acupuncture has demonstrated this statement. The research, the scientific improvement on acupuncture should be emphasized.

The phototherapy has been used successfully in some cases of cancer in China. If our life, our body is built up with photons it is not a fantasy that correct dosage of light and adequate procedure could cure some or even all of the illness.

According to Photongenesis theory life is originated on Earth and only on Earth within the solar system by the action and combination of photons that come from the Sun and the Earth. We are not extraterrestrial come from other planet or any other place of Milky Way. Life is autochthon of Earth because the combination of suitable temperature, climate, water, air and other physical and chemical conditions. But most of all the suitable electric and magnetic forces of the Original Energy that made possible the transformation of photon to low frequency biophoton that concede the auto formation and the power of evolution to each specie of living being at the right time and at the right place which is the Earth. It does not mean in other solar system, other galaxy life could not exist.

Life is the continue absorption, combination, transformation, emission of energy of photons and electrons. Hence, evolution is step by step, species by species transformation of photons to different types and ranges of biophotons!

SYNGULARITY

Energy could transform to matter and matter could convert to energy. Logically the tangible matter of the universe arose from energy, the Original Energy and only could arise from that specific energy, which in a small proportion was transformed to photogens and matter. The formula of Einstein would not make sense if it was not that way, since the energy is equal to the mass times square the speed of light ($E=mc^2$). Before the formation of the universe the mass did not exist, the speed was zero. Logically the equation was equal to zero. Until today the universe should not exist, just because mass and its speed did not exist before!

The same formula demonstrates the Energy O (EO) gave up the exploited, accelerated mass. Then we got to modified a little bit the formula: $Eo = mc^2$. It reflexes only part of the energy O, the energy that had been transformed to matter.

The Big Bang theory stated that the universe was originated from an extremely dense, heavy, hot primeval atom which was the singularity. That mass $m=E/c^2$ was inert, a steady, extremely solid atom, lack of acceleration. Hence, the equation equal to zero. On the other hand, after the Big Bang, there were no mass fragments; it passed through unknown processes up to light elements, radiations and then mass appeared. Why big fragments of mass did not appear after that huge explosion and what force made such dense, extremely compact massive object formed by heavy elements and complex molecules to explode!

The same formula demonstrated that it was the Original Energy EO that produced that exploded and accelerated mass.

During the initial formation of the universe, through Photongenesis processes the entire space was saturated by the Original Energy photogens which traveled and accelerated with the speed of light. In a specific moment: high temperature and pressure, violent vibration and ionization were transformed to plasma. Should other part of Energy O, and precisely that part be the "dark energy" and "dark matter" that everybody talks about but nobody knows what it really is? Isn't it the double personality or the duality of Energy O in its different states? So, if the Original energy is EO, the photogens plasmatic energy that converted to mass is Ep, dark energy Ed and the dark matter Md. The energy Ep that transformed to matter would be: Ep= EO – Ed – Md which represents four percent of the total energy that converted to mass.

The Cosmological Model, the Big Bang theory stated this four percent built up all the heavenly bodies of the universe which come from the dense, hot primeval atom, the singularity. The singularity defines the origin, the beginning of the universe.

The General Relativity theory descripts two types of geometric singularities: the generalized singularity of the entire universe and the focalized black hole singularity.

The singularity as the initial of the universe is based on: the expansion of the space between the galaxies; the presence of light elements at the initial new born universe; the thermal body that left relic background microwaves radiation in the universe; the homogeneity and isotropy which is the Cosmological Principle.

According to the Big Bang theory the universe departed from an infinite dense primeval atom where the gravitational force is infinite; where the entropy is infinite without return. It means the universe was created from the singularity. How could something huge, as big as the universe that comprises uncountable stars and galaxies be created from infinite entropy without return?

The singularity as the classic description that compresses all the matter, space-time and energy of the universe to a point like atom is a myth. There is no single force or the combination of four forces together could constrict a single celestial body, even less could compress the entire material universe. Matter under extreme high temperature in any state, solid, liquid, gaseous or plasmatic would increase its volume several times instead to be compressed. Before it is compressed and under high pressure it would explode, increasing its volume! If that phenomenon is possible, it should be already observable in the space. Hence, before the Big Bang there was no classic singularity. We have already analyzed black hole formation, evolution, development and concluded that black hole is a recycle system not an endless compressing singularity. Therefore the second kind of black hole singularity is also a myth!

The new born universe was a fire ball without matter, the matter dependent gravitational force appeared after the formation of subatomic particles, million years after the Big Bang and it was extremely weak. Hence, there was no singularity event before or at the first moment of the Big Bang. Hence, Big Bang was just a change of phase sometime, somewhere, somehow of the evolution of the young universe.

The homogeneity implies that the universe seems to be equal in all directions, it does not matter whoever or wherever is the observer; it implies the physic laws are in force everywhere of the universe; it implies that any region, any heavenly body, in any epoch could have its individuality but over whole it still form part of the homogeneity.

The isotropy implicates that the temperature and radiations are kept equally distributed in all direction of the universe.

The Original Energy Theory stated that the Cosmologic Principle is rather due to the entire existences of the universe derived through Photongenesis; derived from the same elements which were ultra energetic photons; derived from the same origin which was the embryonic energy formation. They were stretched, emitted from the same fire ball and during all the evolution of the universe which axis rotated mixing everything since the beginning, up to today and forever. If the universe really derived from an extreme dense, hot, primeval atom, the singularity, this phenomenon would not happen.

If one day the fuel of the Sun, hydrogen and helium are exhaust, the nuclear reaction fusion would stop, the Sun would expand. The heat would make the planets one by one to explode until the entire solar system blow up. That phenomenon would be completely opposite to the singularity.

On the other hand, while the Sun alive its masses are consuming and converting to energy and form part of the energysphere. When the Sun is out of fuel its weight has reduced, the weight of all planets, moons and asteroids together might be higher than the Sun´s weight. How could the gravitational force of the Sun or any of the planets cause the singularity since while the Sun in life the gravitational force could not do that? By the same reason, it would not happen in the Milky Way galaxy since while the galaxy alive, possessing extreme strong gravitational force, singularity did not happen, less possible when it is skinny out of fuel.

If the pan singularity should happen, the universe should have a heavy central nucleus and a denser populated center which could make the attraction and contraction from the peripheral to the center of the Universe. That nucleus would stop or impede the spacing out of the galaxies, therefore, also the expansion of the Universe. Long ago the singularity should have happened which is not the reality. The universe keeps expanding and the expansion is even accelerating!

The gravitational force is islet, clumping force that each one or each group of celestial body possesses, making the pan singularity impossible. Only the transformation procedure of the entire material universe converts to energy and then comprise the wavelength of the electromagnetic field infinitely could make possible the singularity.

Now, we know why the heavenly bodies, the entire solar system, the galaxies, even the cluster of galaxies could maintain suspended in a specific place of the space-time, without precipitate or mix up in the macrocosms. Why the universe is so ordered and logic? That is because all the material existing is suspended by the energysphere, by photogens potential time energy, by the structural mesh of the Universe; while they are traveling, expanding with the kinetic time energy, with the Original Energy which keeps transforming and extending the wavelength.

And who govern outside the material space, where there is no matter, no gravitational force? Beyond of any heavenly body, beyond the material visible universe, where there is a vast zone of energysphere. Categorically it is the Original Energy that governs the energysphere outside the material universe.

There is no gravitational force of single or group of moon, planet, solar system or galaxy system by itself or all together could maintain the entire Universe as a whole.

By the same reason because the gravitational force is not the force that governs the universe, not the force that holds all the material universe, not the force that revolves the universe, the singularity as it has been descript and defined would not happen. If some day the universal singularity occurs, the visible matter, the invisible matter, plasma and energy of the universe would be mathematically converted to the equivalent of Original Energy. It would not be caused by the effect of the gravitational force, since as soon as the matter disappears, the gravitational force disappears also. Foremost the gravitational force is the weakest force.

It means, if the pan singularity should happen, any mass, any physical material would convert to atom. The violent vibration would disorder all the electromagnetic superposition of atoms. The temperature would arise to trillion of trillion degree. The electrons would spill out of the atoms; the nucleus of the atom would convert to subatomic particles, then the subatomic particles to photons. The entire material universe with its potential space-time energy would transform to kinetic time energy which is photons with the speed of light. The universe would be converted to a fire ball where only photons exist again!

Photons are the association of electron and positron; most of them would separate or annihilate each other but would not generate photons again as the Photongenesis duplication during the initial of the universe.

Instead, the survivor electrons would joint to the survivor photons energizing them. All the photons of the electromagnetic spectrum would become to gamma rays. More electrons would annex to photons energizing them, becoming to photogens which would reduce their wavelength infinitely becoming Original Energy. During the reduction of the wavelength the temperature would decrease in each step; the electric and magnetic components of the photons would not be perpendicular each other, they would lose their velocity. The electromagnetic field of photons would reduce up to almost disappear. Few amounts of survived photons would become to extremely energetic points, becoming angels without wings, without electromagnetic field. This process where the electrons join to and energize photons reducing their population is an opposite process to Photongenesis; we might name it as photonsynthesis. This process consumes a lot of energy and radiation, because photons absorb all the existent energy of the universe which is electrons. They would be used to form higher and higher energetic photons, even the lower frequency photons would be used to energize others photons up to extremely ultra energetic Original Energy photons.

Because the entropy, the temperature would decrease progressively. When all the photons are on rest without vibration temperature would drop down, done to close to cero K. Every activity would stop. The entire universe would reduce infinitely to compact wavelength of Original Energy and frozen. That would be the real Original Singularity.

FATE OF THE UNIVERSE

The Original Energy theory has postulated the origin of the universe is derived from Photogenesis or Photongenesis procedure, from an energy formation of the Original Energy where the embryonic photons were on rest without any electric or magnetic activity. Consequently, embryonic photons were without electromagnetic field. Space-time inheres with the energy O; the universe was only a mass less, photons energy formation.

At the beginning, through Photongenesis procedure, photons-electron mutual generation, photogens population increased, the temperature increased. The energy commenced to activate. The photons magnetic axis commenced to rotate, inducing the electric components to vibrate. Potential energy of the embryonic photons commenced to transform to kinetic energy. The electric and magnetic components become to perpendicular. Immeasurably energetic, compact, mature photons unfold, stretch, extend, inflating and expanding space-time. Photons flied out with the speed higher than the speed of light. The universe begun to form! Base on this statement and the law of energy conservation which states the energy never disappears only transforms, it is not difficult to forecast the fate of the universe.

Any begin has an end, as a material existence our universe would finish one day. But Original Energy is infinitely vast, therefore space time. The recycle procedure through black holes continually give back energy to the universe, the cyclic transformation is almost endless. Hence, the universe seems to have longevity!

It has been pointed out that the Original Energy forms the axis of every heavenly body; by stretching the wavelength it forms the structural mesh of the universe; it is the force that have originated, extended and expanded the universe. Photogen is the transformer, the force that forms the nucleus of every material existence; it is the spring that forms the mass, the force that make possible to recycle matter to energy; it is the force that holds every heavenly body together constituting solar systems, galaxy, cluster of galaxy and super cluster galaxy system. Energy O and photogen occupy ninety six per cent of the entire universe; they are far away from the emptiness. Hence, the universe is far away to the end!

The Original Energy theory states our universe is an open system, even if the action of entropy occurs it would not finish in a chose, because the continue abundance supply and recycle of the Original Energy. In any system when heat is converted to another form of energy such as electric or magnetic energy, there is always loss of energy, but the heat of every heavenly body is insulated by vacuums, the loss is limited. Consequently, the universe is kept in an amazing healthy, stable, constant development system.

But if the energy continues to transform to matter and the material existence overcomes, the material predominated universe anyway would finish one day. But energy is keeping predominating, conserving, transforming and recycling. That is the perpetuity of the Original Energy!

Any waves need a medium to propagate: sea waves are propagated through water, sound is propagated through air, TV signs waves could propagate through cable network. Electromagnetic waves propagate through the structural mesh of the universe; even people could make waves in stadium spectacles. How the photon waves propagate? The cosmic background microwaves radiation demonstrated photons were, are and still propagated through the electromagnetic field, through the Original Energy structural mesh. Radiations are propagating by itself in vacuum. The cosmic background microwaves radiation also proved the wavelength has been extended progressively. These extensions should have a limit where the wavelength could not stretch anymore. Hence, if the universe was born, inflated, expanded, develops, evolves and would reach its end, the waves should retract and go backward, reducing their wavelength infinitely until a dimensionless point and stop. The universe would suffer a nucleolysis and photonsynthesis reduction, an opposite transformation. Temperature would increase extremely but later entropy would overcome and cool down until close to cero K, when electric rotation and magnetic wave's vibration stop; when atoms converted to energy and stop to vibrate. The electromagnetic field mesh would retract and become a small energy formation.

Meanwhile, most likely the universe would just keep transforming for an unimaginable long time giving a still constant development impression. It is predictable that more and more matter would transform from the Original Energy converting to more heavenly bodies. At the same time they get old and out of fuel. The black holes would continue to recycle all the old material objects. So on, more new galaxies and stars would arise, closing the endless transformation. The universe would continue to expand and would not be saturated because the ultra energetic waves are still stretching and the space is unlimited. The static state would never happen, because the existence of the regulatory recycles system and because there are enormous immeasurable amount of energy O that still could transform.

On the contrary, if the universe arose from a super concentrated matter as the Big Bang theory stated, the first state of the new born universe should be constituted by fragments of matter, not by kinetic energy as it has proved. The static state should happen after the inflation period, because there is no more matter to be transformed at all. The universe should not expand anymore; even part of the celestial bodies would convert to energy. Super heavy black holes would dominate everywhere. The universe should finish soon.

The Big Bang theory has been established the fate depending on the density of matter and the expanding rate; it means depend on the gravitational constricting force and the Cosmic Constant which is the gravitational "repulsive force". The Big Bang universe could have three possibilities to end: Big Rips, Big Crunch and Big Freeze. The fourth possibility is the cyclic transformation but would be the repetition of the singularity which is included in the three possibilities that had been mention. Nobody knows which would be!

In an absolute no space, no time, nonexistence cosmos, outside of our material universe, the most difficult to sustain and even to imagine for the Big Bang theory is: when, where, why and how the material universe finishes? The universe is inside a close hermetic capsule? The Big Bang universe should finish where the oldest stars have reached. Does it indicate the universe has reached its limit or it would still keep growing? Is there anything outside that limit or outside the material universe where is no space? If the universe keeps expanding forever the heavenly bodies would separate more and more. How could the each time weaker gravitational force cause the singularity?

There are no convincing answers.

The Original Energy theory states that in the case the pan singularity should happen or if the universe should finish, Photonsynthesis procedure would enter in action, explosions of all material heavenly bodies should happen everywhere at the entire universe. All the matter formation should disintegrate inside black holes first. Then all the black holes confluent in one big black hole, convert all the matter into energy. Temperature would arise trillion, trillion and trillions of degrees while all the matter burned up; the energy of the entire universe would squeeze and be compacted; wavelength become shorter and shorter up to infinite. Electrons would be disseminated all around the peripheral of the universe and transfer the energy to photons. All the common photons would be compress more and more converted to gamma rays, then to extra gamma rays which are photogens and then to ultra gamma rays to Original Energy. Photons population would reduce because all of them are added and energized by electrons becoming extremely compact wavelength photons. Energy is consumed, temperature would decrease.

After that event, the extremely hot and dense energy enters to entropy; some free electrons would annihilate with positrons, but would not form partnership with photons to form matter anymore. Photons electric and magnetic loops would reduce, photons go into rest. Temperature would drop down up to cero K. Space-time collapse and vanishes. Rotation stops, electromagnetic field reduced to almost cero. Only Original Energy would persist in a hibernation state until a new transformation. New universe would reborn.

The Original Energy theory postulates, the actual universe is an opened universe. Our universe is several times bigger than the one it has been calculated because its energy reaches beyond to the oldest galaxies and other object have traveled; it means the extension of the universe could be more than three hundreds billions of light years. So the Original Energy universe might keep transforming; just a tiny portion of the Original Energy had been used. The wavelength of all kind of ultra and extra energetic electromagnetic photons could keep extending and occupying more space until they could not stretch any more. But thanks the recycle system of the universe and the energy conservation, the universe would keep its longevity.

There might be others universes around and then our universe would reach its limit. Foremost from the same Original Energy could derive multiple universes and all of them keep expanding, dividing or multiplying. But all of material universe would end, recycle and be reborn. None material formation would last up to the eternity as physical existing. Only the transformation of the Original Energy is endless.

The Original Energy last forever!

FATE OF HUMANITY

Nature is the way the Earth, the solar system, the galaxies, the universe and everything go naturally. It does not mean nature always elapses smoothly. Nature takes action with enormous energy; nature establishes the equilibrium (order, isotropy, homogeneity, diversity) with extreme force. We experience it with Earthquake, tsunami, volcano, diluge, flood, huracanes, tornado, el niño, la niña etc. . . . but it is the force necessary to change the alterations; it is the force that naturally makes possible the transformation; it could be one of the multiple stages of the natural evolution. Would the intelligent human kind permit the nature keep going naturally? Or the intelligent human kind rather conquers the nature altering it? This is the most likely the fate of the humanity, the auto destruction by altering the nature, the environment, the diversity, the atmosphere!

Soon or late the motherland would be saturated, would be over exploited. The natural resources would exhaust. Human kind would pray the heaven for more bread and water, human kind would beg for what nobody never care, never worry about: the environment, the diversity! When human kind realize we cannot exist along, when all around has been consumed, destroyed! When human kind eats each other!

May one day the humble reflection permits the human kind to realize that the environment should be respected, should be protected, recreated and even should be repaired?

We depend on each other; the diversity should be permanent, should be permitted and should be protected.

The body, the brain are objective matter; the conscious mind and other cerebral activities are subjective energy. Both have incorporated to make a functional unity.

The heavenly bodies are objective matter; the four forces are subjective energy. Both have incorporated to make a functional inseparable unity. But all are limited in their unit.

The neurons, the endothelium of the heart and arteries, the endothelium of the cornea are life last structures, not substitutable, not regenerateble. But they engage their complex function as one day all the time. Foremost, the genes, the neurons that constitute the living being since the most primitive bacteria up to the humankind are extremely similar. The genes only vary one to few percent between them. The same atom, the same composites have been used to form any kind existing on Earth, on the solar system, even in other galaxies. The same molecule of oxygen has been breathed trillions of times. What makes the difference is the amount of Original Energy the structures contain and the constant transformation of that energy on those structures are capable to make.

The original energy not only gives origin to every existing but also it situates them in a surrounding space and time, determining their existence in the microcosm and macrocosm. Consequently we should worry about this unity, this integrity, this harmonic coexistence.

We have the privilege to live on the Earth planet and be the conscious mankind. Ironically we do not know with what we are made?! Any and every living activity on Earth comes from the light and heat of the Sun. it means comes from the electromagnetic waves, comes from the photons. By Photongenesis, nucleogenesis and nucleosynthesis the photons have been converted into biophotons. The biophotons build all bodies and soul of all living beings including human kind!

Thus we are made up with waves, with photons! We bodies and souls are photons. Just like the entire universe.

During more than four and a half billions of years, the Earth electrically, magnetically, chemically physically and biologically has transformed uncounted of times, just like the weather and climate. During this time, living beings have being born, evolved, developed, dead, replaced generation to generation, and species to species, a lot of them have disappear forever. Haman kinds could have the same sorts. The force that makes those transformations is the energy. The Earth would keep transforming, the Sun would keep transforming, so on humankind.

The difference is the electromagnetic radiation is using as telecommunication, medical technology. Even the harmful ranges are using broadly as ever for nuclear power as atomic bomb, hydrogen bomb or thermonuclear plant. "The intelligent mankind" now is capable to use extreme energy contributing to the alteration of the natural transformation, destroying our self, destroying the Earth! Knowledge has not made human be less evil! But the worst of all are the endless wars, the each time more sophisticate, massive destructive war! One day could reach the pan auto destructive genocide.

The inadequate use of radiation could cause cancer. Even the simple cell phones that almost everybody uses, bomber with electromagnetic radiation all day long could interfere with the electromagnetic field of the body, could cause damage in endocrine glands, ovaries, testicles and the brain, and could alter the concentration, memory and neuronal activities.

All the layers of the atmosphere has protected us since our existence, otherwise we could not survive. Paradoxically we alter, destroy them and not care about of their conservation, demonstrating our unlimited ignorance.

Something is radical and drastic: if everything derived from the photon energy, great scale or small accumulative alteration, destruction of the electromagnetic field could certainly for sure, cause huge damage. We barely know the alteration of the ozone layer makes possible the ultra violet radiation reach us and cause cancer on the skin, cataract on the eyes. But the atmosphere is consisted with different layers and all of them contribute to the transformation of ultra and extra gamma rays, the attenuation of solar radiation and the conversion of all harmful rays to lower frequency and longer wavelength photons. The each time worst natural disasters phenomenon is alert signals that those layers are altered.

Hence we should be conscious and stop all activities that could alter and damage the atmosphere, our life protector!

The fate of human kind, depend on precisely the conduct, the behavior and how intelligently keep using the energy which has not been too bright up to today!

We have the grand opportunity toward the wide opened horizon, to the completely open sky, to the glory of heaven. What impedes us is our nearsighted mentality, our meanness, our savage nature! The more we know how to use the electromagnetism power, the wilder behavior and more atrocity we are capable to make!

We, human kinds presume we are intelligent; we deserve everything on Earth, but we waste and destroyed everything; we kill each other to possess the Earth which barely is a tiny corpuscle in the immense universe; our meanness is as big as the space! Human kind has not realized: space is our heaven, space is our future and space is part of our home.

The Original Energy theory might offer a lightening. Be true or be fallacy the Original Energy Theory, the author tries to open human's eyes and minds. Our opportunity is immense, just like space; the resources are immense, why to fight? Why to hate! Why to kill! Why create different Gods and impose others to believe? Why annihilate each other?!

Are we electron and positron and be annihilated each other?

We are photons we should travel ahead like rays; we should be intelligent being as photons!

Sky is not just up. Heaven is all around the Earth, all around us, Original Energy is all around us and they are ONE!

EASY, IMPOSSIBLE AT THE SAME TIME

The Photongenesis theory establishes that photon was the first being in the universe; every existence in the universe is derived from photon. The Original Energy is the structural mesh of the universe through which the photons propagate; it orders, organizes and rules the universe.

If I got to prove the existence of the Original Energy and the about statement, it is easy and impossible at the same time. Easy because all I have already expounded before, because the truthful of the Big Bang theory is doubtful. It is impossible because it is difficult to accept that the energy at the extreme inside of every nucleus, the core of the galaxy and others celestial bodies are pure Original Energy. It is very difficult to accept that our soul is made by Original Energy and even more difficult to accept the Original Energy is the ruler of the universe. No devise could prove that. But we could feel and observe their effects once the transformation of the energy takes place revealing the reality, ones it is materialized. Our own Energy is connected with the electromagnetic field of the Original Energy.

Once the Earth was formed, four and a half billion years ago, the living being also started to form, from an ultramicroscopic what I have called Biogene. Biogene existed and has conserved up to today in every living being. It has been transmitted through the evolution to every individual, to every species. It could be compatible to everyone in transplants, implants procedures without the risk of rejection. These common stems in reality contain the Original Energy constituted with biophotons which is the origin of life.

The cell's exponential multiplication by cellular subdivision in reality is another transformation, a way of Photongenesis of the Original Energy.

Our Sun or others stars are constituted primordially by energy, the extra energetic photogens, combining with electrons they form the electric repulsive force that erupts out of the core which is the real source of the solar radiation and light of photons. Hydrogen, helium and other elements are only the material manifestation that scientists have detected. The real energy of photons comes from the core, the nucleus.

Maybe in the future the cosmic rays could be measure with more precise accuracy, and then we would be able to prove that gamma rays are not the most energetic rays. The existence of extra energetic rays which are E rays or photogens in this theory is already possible to detect. The ultra gammas rays which are Original Energy, O rays are also have been detected in the space as ultra energetic gamma rays or short wavelength rays and as cosmic radiation bursts. The problem is to detect them in the core of every heavenly body, inside of every seed, inside the nucleus of the reproductive cells, inside of our brain. They would not be detected as ultra energetic photons inside our brain, because they have been transformed and attenuated, but keep their characteristics, capability and power. That is the challenge!

It is possible the extreme high energetic photons not directly convert to common photons of the spectrum straightforwardly but in different ways, different kinds as coherent photons, biophotons and regular electromagnetic spectrum photons.

We are direct products of the attenuated photons of the Sun and the Earth; we are not derived from extra terrestrial meteorites; meteorites and cosmic rays only contributed part of components. Once again: we are derived from the Sun´s and the Earth´s photons, not from cosmic rays. They might have contributed with some ingredients as complementary elements. Thanks God we are not look like extra terrestrials!

In the beginning I mention an "explosion". Maybe I was influenced by the Big Bang theory. But each time I have been convinced myself that the formation of the universe was through a prolonged evolution also, which took place not through thirteen and a half billion years ago but maybe through twenty, thirty billion years or more, accounting the embryonic period. That is because we got to take in consideration the period before the explosion which could be tiny ultra high frequency photons in a freeze and squeeze compact state. Then, by Photongenesis and duplication processes the universe was born.

If the Original Energy Theory still does not convince anybody, please make the last voyage from the point we stand, back to the beginning of the formation of the Earth, of the planets, of the Sun, of the Milky Way galaxy; back to the formation of mass, elements, subatomic particles, quantum, gravity, and quantum mechanics... back to the energy. Visualize our Original Energy formation, our first universe gene, emitting the first wave, the first photon duplicating, increasing their temperature! And then let's go back to the future, passing the whole formation procedure, passing from photons and electrons to form subatomic particles, to form atoms then molecules to mass, and one by one the stars, one by one the galaxies, the nebulas, the clusters of galaxies and keep going beyond the visible material universe, through the universe energysphere zone, up to the limit of the universe.

Now, come back to the point where we departed. What have we done? We traveled through the entire universe. We reviewed the formation of the universe. We saw every event, step by step of the transformation from the energy to the matter, to the end of material universe, to the real end of energysphere, the limit of the universe.

How we did it? We did it with our imagination. We traveled a distance of more than thirty billion years, maybe a distance of three hundred billion light years or more. What had taken more than fifteen billion years in its formation we did it in fifteen minutes!

Then our minds travel faster than the speed of light! Do we have Original Energy inside of our brain?

Suppose we are at the border line of the universe, the lighted and lightless edge, the frontier to the absolute darkness. We will be attracted back to the material universe by the gravitational force, if the Big Bang theory is truthful. But we would not be; we would feel we were traveling faster than the speed of light, escaping from that attraction, flying beyond the expanding universe! Or more precisely, flying with the Original Energy at the borderline!

And what happen if we surpassed the limit and enter to the absolute darkness? Enter to the absolute emptiness? We difficultly could imagine what happen on that side. The imagination suddenly stops. We would not be able to make any speculations with certainty. We only could go with the Original Energy not surpass and be beyond of it. Then, nothing travels faster than the Original Energy!

It is unnecessary to prove the existence of all range of electromagnetic spectrums, they constitute our every day environment; to prove the existence of higher frequency as extra gamma rays which has been named as photogen undergoes certain difficulty, but the existence of gamma bursts, the extra energetic gamma ray in the cosmic radiation and the formation of high energetic photon inside the black holes could prove their existence. Nevertheless their function as transformer has not proved. On the other hand, the extremely high ultra energetic rays have been extended, unfolded for more than fourteen billions of years, they have transformed to all kinds of less energetic photons.

What is really difficult to prove is the existence of the Original Energy, the trillions electro volt ultra energetic gamma rays. Even the ultra energetic gamma rays could be found in the cosmos, their function as the Photongenesis theory stated is extremely difficult. The most difficult aspect is the Original Energy is present in all the nucleus of heavenly bodies, cells, ovum, spermatozoids and neurons. But the universe is so ordered, homogeneous, logic, that the Original Energy proves by itself its existence. There is an undeniable energy, a perceivable force that rules us, that rules everything, that rules the universe!

We could make some reflections: the small Monarch butterflies fly from Canada to Mexico every winter. If they only consume what they eat, they won´t be able even to arrive to Iowa. But they have the genes which are Original Energy that rules them, guide them, support them, and urge them every year. This is not an instinct!

Instinct is our habit to attribute everything that we don´t understand or ignore about animal, as animal instinct!

For sure everybody has asked: where the children's energy comes from? They play, cry, fight all day long and don´t stop! Some of them might make you crazy; you got to order them to be quiet, to behave. That is the Original Energy the children born with it! And what happen if we put attention on us, we are tired, we always complain about lost of memory, lost of ability . . . . Exhausting the Original Energy!

I might be accused as devil's advocate because I stated that photon is the most fundamental constituent of every existence of the universe; photons are the origin of life, the origin of every existence in the universe! The Original Energy is the ruler of every existence, the ruler of the universe!

Scientist's effort and their discovery not only have contributed to the revelation of the reality of the cosmos; they have contributed to the comprehension of the heaven; they also have made the pragmatic, subjective theology more understandable, explicable. We can go all around the churches, the museums, all around the world to see the pictures, the mural painting and understand what the heaven has been descript. We would never find any as the Hubble telescope has showed.

Being theologies, Copernicus, Kepler, Galileo, LeMaître pioneer of the scientific cosmology, they dared to correct the mysticism, contributing to the reality.

Lethargic time used to past slowly during millennium and centuries, Galileo, Newton made time to accelerated changing the living style of people in the entire world!

We don´t have Gods of everything to explain the nature phenomenon; we don´t pray those Gods to have better harvest, better quality of live.

The existence of God is still undisputable, beyond doubt! Would one day religion and science coincide in one point?

God is light, God's power is energy; conscious scientific reasoning is energy. The Original Energy is energy, energy of photons and photon is light.

CONCLUSION

The Original Energy theory postulates that the Universe is originated from cold, fold, compressed energy formation of embryonic photons; through Photongenesis procedure, a mutual generation between photons and electrons from ultra energetic photons to less energetic photons. Photons derived, constituted every existence of the universe including life. Hence, the Original Energy theory is the energy that forms, derives, develops, governs, transforms and rules the universe. It means the universe did not derived from infinitely hot, dense matter where the mass dependent gravitational force was the primordial force.

The Original Energy theory establishes that the universe derived from the embryonic energy formation which was cold, containing ultra energetic photons with the energy trillions of trillions of times stronger than the actual photon´s energy. They possessed all the codes capable to generate all existence inside the universe.

The Original Energy Theory is an energy based theory, which affirms that every heavenly body, object or energy formation possesses intrinsic ultra compact Original Energy inside the core, responsible of every rotation, formation, action, reaction, evolution and transformation. Being photon the basic unit of the electromagnetic interaction, it implies that such creative energy was electromagnetic energy. It implies that it was not the massive primeval atom and the mass dependent gravitational force from where the universe derived from.

It establishes that before the actual universe was formed, the Original Energy or the Energy O already existed, where the embryonic photons were conglomerated in a tiny compact, freeze energy formation. The Original Energy embryonic photons were extremely small. Under almost 0 K temperatures, photons were points like without rotation or movement, without electric or magnet vibration, without electromagnetic field around them. Hence, the Original Energy was one without distinction as kinetic or potential energy; neither as space time energy; there were no four forces distinction yet.

Time existed as the eternity past, since the infinite past, belonging to others cyclic transformations.

Space existed as a Virtual Space included inside the compressed Original Energy, characterized by the absolute mass emptiness. Space-time was inherent inside the Original Energy, all in one energy concentration.

During this embryonic stage of the universe, mass was zero, space was zero, time was zero, even the temperature was close to zero. Everything began from zero but the immense energy.

When the embryonic photons mature, the Original Energy through Photogenesis processes emitted messenger photogens, precursors of the actual photons with infinite compact wavelength, capable to develop extreme high frequency.

From photogens derived electron and positron which reacted converting to pairs of less energetic photogens; photogens generated pairs of electron and positron which were converted to pairs of less energetic photogens again; from new photogens derived new electron, positron emerged, transforming to less energetic photons. They rotated, vibrated and polarized; the magnet bar was formed inducing the electric formation. The electromagnetic field was completed. This Photogenesis, a mutual creation between photons and electrons processes was repeated again and again.

When there were excessive amount of photogens inside the compact system, they vibrated intensely, the electric and magnetic components were extended and turned to perpendicular to each other. They commenced to unwind and stretch violently elevating extreme high temperature and pressure producing a huge explosion. Photons flied out faster than the speed of light!

The dense, compact Original Energy's waves vibrated, stretched, extended, unfolded violently, establishing the electromagnetic structural field. Space and time appeared with the inflation forming the energysphere. Original Energy, photogens, photons, electrons and neutrinos occupy the universe as an isotropic and homogeneous fire ball during a long period of million years.

From Photongenesis procedure we can observe photon could transform to electron and positron, during the annihilation photon becomes to neutral charge and the mass is consumed in the annihilation. That is why photon is no charge and no mass. But when photon is transformed to electron, electron acquires a tiny mass. That is the secret of the transformation from energy to matter and from matter to energy which is the secret of the uncertainty.

Since then and forever, from photons derive kinetic time energy and potential space time energy. Photon is built by kinetic and potential energy. During the energy dominate epoch, during the high energetic photon fire ball epoch, there was no gravitational force. Photons were the elements that filled up the entire universe making it isotropic and homogeneous.

After series of explosions millions of small fire balls were disseminated, altering the isotropic and homogenous distribution of the temperature and energy. The universe enters to the Nucleogenesis process.

In a further stage, the annihilation produced nuclear reaction in the nucleus of future galaxies and stars formation, establishing the nucleosynthesis procedure, deriving gamma rays, quacks, gluons, neutrinos and others subatomic particles. The baryonic matter appeared.

Once the subatomic particles were formed, gravitational force emerged, but it was too weak, too late and too slow. The positive pressure radiations overcome. Hence that is the way matter overcomes to antimatter; electromagnetic force overcomes gravitational force; that is why the new born universe could formed!

Original Energy theory states that all activity of the universe derives from energy. Fusion is realized in the core of our Sun, where high energetic photogens releases radiations with positive pressure, which overcomes the attractive gravitational force. This radiation pressure also defeats the well known electron-positron annihilation, releasing less energetic photons.

Base on these statements, it is postulated that the Original Energy rules and constitute the universe. The electroweak force, strong force and gravitational force derived from the electromagnetic force which is the fundamental constituent of the Original Energy and of the universe. We also could conclude that energy is the fifth state of the universe beside solid, liquid, gaseous and plasmatic state.

The Original Energy theory has postulated that the energysphere that radial from the core of every star, every galaxy, every heavenly body is formed with the combination of photons and electrons. That way photons and electrons form the electromagnetic field, form the gravitational force and spacetime! To prove and demonstrate how they combine to make all these work is the highest challenge.

The electron is the smallest, lightest particle and as fundamental as the photon. During the initial Photongenesis process each photogen generates a pair of electron and positron which reacted becoming to a pair of less energetic photons. These new photons were converted to electron and positron, then they converted to a pair of less energetic photons; photons transformed to electrons and positrons again. These endless phenomenons occur inside the nucleus of stars and galaxies up to today. What implies the wavelength got longer and longer whiles the frequency become weaker and weaker.

The magic occurs right here in the transformation of photons which is energy, to electrons which is matter. The secret resides in the transformation of kinetic energy to potential energy and from potential energy to kinetic energy; from invisible energy to visible energy.

The entire secret of the origin of the universe; the origin of every existent inside the universe; the origin of life are enclosed in the transformation of the most tiny elements photons and electrons, from energy to matter and from matter to energy!

The entire universe was originated from the same ingredient which is photon. Consequently, everything obeys the same physics, chemicals, electric, magnetic, biologic laws. Everything is similar and homogeneous in the universe no matter whoever or wherever is observed; the temperature and the microwaves background radiations are isotropic because inside the universe everything come from the same origin which was the photon's fire ball.

The Original Energy constitutes the nucleus and axis of any formation of object or energy. It rotates giving disc, elliptic, spiral, spherical shape or any other form to every heavenly object, enclosing them inside of an energysphere. Hence, the radius of the sphericity of these objects or energy formations should be considered as the fifth dimension beside the width, long, high and time dimensions.

This theory established also that any heavenly body or object of the universe has its energysphere that comes from the core and surround it. For instant: around the nucleus of the atom, its cloud of electrons and their energy forms an energysphere; the Earth possesses its geoenergysphere; the Sun energy forms solar energysphere; the solar radiations spread further away the planets and reaches beyond heliosphere, heliopause up to the bow front.

According to the Original Energy theory any heavenly body possess energysphere emanated from the core; matter and energysphere form an inseparable unity. Together form the attractive gravitational force which is the interaction between photons and electrons. It means, the gravitational force, not only depends on the matter as Newton affirmed, neither depends on only the curvature of spacetime as Einstein argued. The energysphere is kinetic energy that has been transformed from the core and increase until the star or any material is out of fuel.

Vacuum is not empty it is constituted by the Original Energy mesh and the transformer photogenic energy which constitute the most important part of the universe. The energysphere is the place of action, interaction, confrontation and division between the heavenly bodies where the energy interlaced. That is why every heavenly body could be "suspended" and keep rotating in its orbit, in the space.

The energysphere of the Sun acts as a convex lens causing a mirror effect, making the hidden background far stars image could be seen in foreground, when the photons of the star are deflected by the energysphere of the Sun.

When far away light source (light from galaxy, star or black hole), the Sun and the observer, these three elements are completely aligned, the energysphere appears as a ring around the Sun, because the light from background put in evidence the energysphere of the middle situated Sun which acts as convex lens refracting the far star light.

We can state the universe has an energysphere that reaches far away beyond all the galaxies, beyond the old galaxies and any heavenly objects have reached, which could be several times the size of the actual observable universe that has been calculated. Hence the age and the size of the universe also should be several times the actual age or size that have been stated.

Einstein recognized the Cosmological Constant was an error but actually it has been revived more and more because it is necessary to explain the fate of the universe; because it is necessary to explain the existence of the "repulsive force" that causes the expansion of the universe for the Big Bang theory and the Cosmological Model.

The Original Energy theory states that the expansion does not depend on such "repulsive force", it depends on the rate of the extension of the electromagnetic waves and the constant transformation of the kinetic energy to potential energy and vice versa. The universe would keep longevity; the Original Energy would last forever.

The speed of light is constant in vacuum, but is inversely proportional to the density of mass; for the same reason, mass does not travel with the speed of light as it has been established!

The galaxies and cluster of galaxies are hold together by the photogen energy which is potential time energy. On the other hand, the black holes recycle system retracts compresses and recycle continually all the stars and galaxies that are exhaust of energy. Hence, while the electromagnetic kinetic time energy of the Original Energy makes the universe expands; the potential time energy of the photogen holds the material universe together, and the black holes constitute the recycle system. That is the most fundamental mechanism that keeps the universe ordered and stable.

If we analyze the volume of an atom, the nucleon is extremely small; the electron that rotates around is even smaller. They occupy an insignificant space. The rotation of the electron is electric charge creating electric field and the electric field induces the formation of magnetic field; together form the electromagnetic field. The rest of the volume of the atom is empty space fill up with electromagnetic kinetic energy! The mass of an atom might be only 4 % of the entire volume of the atom. This phenomenon occurs in the entire universe. It means the principal ingredient of the universe is the electromagnetic energy! The universe keeps stable, isotropic and homogeneous, almost static because this 4% is enough. The universe does not need more mass to keep its existence, is the electromagnetic structural network, the energysphere of the Original Energy that form the universe primordially.

Then, the concepts of dark energy, dark matter, Cosmic Constant and singularity need to be reviewed.

The Original Energy Theory has postulated that cerebral mental activities as mind, intelligence, memories, consciousness, reasoning, dreams, thought, intuition etc. are made up with waves of biophotons derived from the Original Energy.

The Photongenesis theory states the entire existence of the universe derive from photons. By the same reason, life as plants, animals and intelligent being were originated from the Photongenesis procedure of photons. The solar radiation which is the spring of everything is photons, they acts on organic molecules converting them onto primeval unicellular plants which had the faculty to realize photosynthesis. After that, the unicellular bacteria were formed. Unicellular plant and unicellular animal formed the symbioses environment.

This implies the mutual supply of oxygen and carbon dioxide and even more important the administration of energy ATP. That was and is the way biophotons, highly ordered and synchronized, low frequency photons appeared and that is the origin of life!

The ability as messenger of photon makes possible the apparition of the vital messenger ARN, which makes possible the transference of every life codes. The formation of ADN makes possible the transference of genes from generation to generation. Ones again, any transformation from simple light chemical element to heavy element, from elements to composite, from composite to molecules, from inorganic to organic needs the energy of photon, needs the messages of photons, needs the partnership combination of the mutual generation of photon and electron. All of them are the result of photogenesis. These are the most fundamental mechanism of the formation, transformation and evolution.

Life comes from the Sun's photons and electrons combine with photons of the electromagnetic field of the Earth which is situated in the comfort zone of the solar system. Hence, photons biological transformation to biophotons on Earth is the origin of life.

Meteorites that reach the Earth suffer collisions with the cosmic radiations, with different layers of the atmosphere of the Earth and crash with the surface of the Earth elevating extreme temperature. Most of them disintegrate and evaporate by friction and collision. Consequently, life if any that comes with meteorites would not survive!

Life comes from light, naturally from the Sun; life was originated on Earth principally by the transformation of photon and electron to week energetic biophotons. Then by the combination of biophotons and the specific, intrinsic conditions of the Earth! One of the most outstanding factors is the electromagnetic field of the Earth which not only protects the Earth against the harmful radiation but also supplies the suitable photons to become biophotons.

Once galaxies and stars are formed matter decays, atoms give off radiation. Electrons jump out from more energetic orbit to less energetic orbit releasing less energetic photons. Inside the nucleus of the galaxy or star the nuclear fusions release photons by the same way from higher energetic photogens to lower energetic level of photons.

Hence, the entire universe is made up from the transaction of electrons and photons and its final product from disintegration is always photons. Everything is photon, from Photongenesis up to Photonsynthesis activity, from the initial up to the end of all complex transformations. Inside the universe, life is from photon to photon, not from dust to dust, there is no other secret!

The Original Energy or Energy O derives, rules, orders any existence of the universe. This is the similitude between science and theology.

The Original Energy theory or The Photongenesis theory differs substantially to the actual predominant The Big Bang, the Cosmological Model or the Standard Model theories. Some questions that could not be explained by the Big Bang theory now could be explained by the Original Energy theory.

The Original Energy theory establishes that the Original Energy rules, revolves, evolves, expands, recycles and maintains the whole universe as a live, functional unit.

Under the guidance of the Original Energy, Photons gave origin to every existence in the universe. The origin of all heavenly bodies, all matter and the origin of life are attributed to the most fundamental element the Photon under the Photongenesis process. Photogens possess all the codes of every existence in the universe, which are uncountable varieties of different combinations of electromagnetic waves. Photongenesis theory affirms nothing come into existence by random unguided process.

The universe is not created from nothingness; nothingness only could create nothing!

REFERENCES

Johnnie T Dennis Complete Idiot's guide Physics (second edition)
Complete Idiot's guide to Astronomy
Brian Greene "The elegant universe"
Richard Hammond PhD "The unknown universe"
Terence Dickenson The Universe and Beyond
Gary F. Moring M.A. The complete Idiot´s guide to
Theories of the Universe
Carlos Nóbrega, Biophoton the language of cells.
Kok-Haw Kong. Physics,
Stephen W. Hawking The Theory of Everything
George Gamow Gravity
http//en.wikipedia.org/Wiki/Dark Matter.
http//en.wikipedia.org/Wiki/Universe
http//en.wikipedia.org/Wiki/Gravity
http//en.wikipedia.org/Wiki/Cosmic microwave background radiation
http//en.wikipedia.org/Wiki/Quantum optics Standard model
http//en.wikipedia.org/Wiki/Photon
http//en.wikipedia.org/Wiki/bioluminescence
http//en.wikipedia.org/Wiki/electron
http//en.wikipedia.org/Wiki/Quantum gravity
http//es.wikipedia.org/Wiki/Energía de punto cero

http//es.wikipedia.org/Wiki/Partícula subatómica
http//es.wikipedia.org/Wiki/Campo gravitatorio
http//es.wikipedia.org/Wiki/Onda
http//es.wikipedia.org/Wiki/Biophoton
http//es.wikipedia.org/Wiki/Big _Bang
http//en.wikipedia.org/Wiki/Dark energy
http//en.wikipedia.org/Wiki/Cosmological Principle
http//es.wikipedia.org/Wiki/Universo
http//en.wikipedia.org/Wiki/lensing
http//en.wikipedia.org/Wiki/gravitational lensing
http//en.wikipedia.org/Wiki/Standard Model
http//en.wikipedia.org/Wiki/Black Hole
http//en.wikipedia.org/Wiki/Galaxy
http//en.wikipedia.org/Wiki/Vacuum Energy
http//en.wikipedia.org/Wiki/Electromagnetic spectrum
http//en.wikipedia.org/Wiki/Space time
http//en.wikipedia.org/Wiki/Temperature
http//en.wikipedia.org/Wiki/Electron
http//en.wikipedia.org/Wiki/Quantum mechanic
Stephen Hawking A Briefer History of Time
http//tuberose.com
Dr. Riejo Makela Cells are electromagnetic unit
Dr. Chaim Tejman Gran Unified Theory
www.emc.maricopa.edu/faculty/farabee/BIOBK/BioBookNERV.html
The Nerve System
Frank Rubin The theory of the universe. February 17, 2000

ABOUT THE AUTHOR

The author past his entire childhood in the countryside, uncountable nights, he was seduced by the starry sky. What is the meaning of those blinking bright? Until today this hidden mystery persist in his sight.

Drastic family, politic, social changes has been suffered in his life, what obligated him to wander, struggle and fight. Hence, it is not difficult to understand, why he has been a farmer, worker, constructor, writer, physician and ophthalmologist to keep stand.

But as he presumes: he is the last citizen of the world, until today he has not made anything well! He hopes with this book he could contribute a tiny shine on the sky.

Never has been so difficult to acquire knowledge without any career, any help to write about astrophysics! The author apologizes for his ignorance and daring.

He attributes the origin of the universe and the origin of life to the most fundamental element the Photon. Be true or be fallacy the Original Energy theory, the author tries to open human's minds and eyes.

Human kinds should realize space is our heaven, our future and part of our home. The author hopes us to join together peacefully to inhabit the heavenly bodies.

Author of:

"SENDAS DE AMOR"
"ECOS DE REFLEXIONES"
"LUZ ALMA";
"THE ORIGINAL ENERGY THEORY"
"PHOTOGENESIS"

Eight songs have been made in CD by Hilltop Record in Hollywood
Author´s next book: "MELODIES OF ILLUSIONS"

www.ingramcontent.com/pod-product-compliance
Lightning Source LLC
Chambersburg PA
CBHW030753180526
45163CB00003B/1009